gestalten

grid index is a decisive visual dictionary on two-dimensional grids and geometric tilings. not only an 'artwork' in itself, this book can be considered as a reference book for designers, visual artists, architects, researchers, mathematicians or anyone else with an interest in basic structures of the visual world.

the use of grids has always fascinated me. this lively interest, evident in both my visual as well as my sound and installation work, has inspired me to compile a constitutive book on the subject of pure 2d geometry. the aim here is not to present innovative breakthroughs in mathematical research but to assemble, for the first time, centuries of geometry research results in book format. gathering information that can be traced back to plato, islamic tiles, kepler and dürer, or more recently to roger penrose, all this content was scattered throughout multiple sources from scientific papers to artworks or architectural elements. the material included hence belongs to public domain.

overlapping the areas of scientific research and the visualisation of information, this book comprises a very complex array of surface subdivisions, commonly known as grids and tilings. for presentational purposes, these have been ordered from the most simple grids – made up entirely of squares – up to the most complex non-periodic ones showing unpredictable growth patterns and irregular polygons. mathematical rules and formulas were used to guide the construction processes to achieve this systematic classification.

the final result is virtually universal and available for any platform, operating system and graphics software. the files are made available in vector graphic data files on an enclosed disc, which can be used for any purpose.

i would like to thank everyone in the 'studio', especially nuno da luz, who did the research and elaboration for the project.

carsten nicolai, berlin in december 2008

_book indexation system that goes from the most basic regular and uniform geometrical tilings to the most complex irregular ones; a growing array of shapes, vertex transitivity, and symmetry axis present us with an index of plane subdivision possibilities.

grid index is comprised of 294 grids and tilings.
fundamental geometric polygons such as the triangle, square and hexagon serve as its basic building blocks; delving further, one also finds octagons, dodecagons, pentagons and a multitude of rhombi variations.

all grids are provided in vector graphics as well as in printed form.
in the latter, the pages of *grid index* conform to a uniform layout, an archiving format that provides easier access to content: the attributed catalogue number correlates to file name, grid class correlates to folder name, references correlate to bibliographic entries.

this system provides the user with the ability to transform and combine the source files in an almost infinite number of possibilities.

_disc features the 298 source files for each one of the printed patterns.
all listed files are included as scalable vector graphics (.svg) and also as portable data format (.pdf) for preview purposes.
a detailed listing with catalogue number, class, name and references is provided in the next pages, as well as a small content preview.

for further information, please consult *licence.txt* as included on the disc.

_content *grid index* is divided into seven main chapters, each one presenting a further step in the complexification of plane subdivision:
gi-og . orthogonal grids sub-chapters: orthogonal grid union, logarithmic grid.
gi-rt . regular tilings sub-chapter: triangulation.
gi-st . semiregular tilings sub-chapters: semiregular tiling dual, tiling union.
gi-va . tiling variations.
gi-dt . demiregular tilings sub-chapters: k=2 uniform tiling, k=3 uniform tiling, k=4 uniform tiling, k=5 uniform tiling, k=6 uniform tiling, k=7 uniform tiling.
gi-pt . pentagonal tilings
gi-qt . quasiperiodic tilings sub-chapters: 3-fold symmetry, 5-fold symmetry, 7-fold symmetry, 8-fold symmetry, 9-fold symmetry, 12-fold symmetry.
gi-npt . nonperiodic tilings sub-chapters: penrose tiles, 10-fold symmetry.

_appendix further contextual information, bibliographic references and notes.

cat. number	grid class	name and type	references/ notes
001 - 017	orthogonal grid	100 mm	square grid
		50 mm	
		25 mm	
		10 mm	
		5 mm	
		2.5 mm	
		1 mm	
		1 pt	
		1 agate	
		1 pica	
		1 cicéro	
		¼ inch	
		½ inch	
		1 inch	
		2 inch	
		4 inch	
018 - 024	orthogonal grid union	100 mm + 4 inch	
		50 mm + 2 inch	
		25 mm + 1 inch	
		10 mm + ½ inch	
		5 mm + ¼ inch	
		2.5 mm + 1 pica	
		1 agate + 1 pica + 1 cicéro	
		1 mm + 1 pt	
025 - 036	logarithmic grid	2 cycle semi-log	
		3 cycle semi-log	
		4 cycle semi-log	
		5 cycle semi-log	
		6 cycle semi-log	
		7 cycle semi-log	
		2 cycle log-log	
		3 cycle log-log	
		4 cycle log-log	
		5 cycle log-log	
		6 cycle log-log	
		7 cycle log-log	

001 - 003	regular tiling	{3, 6}	isometric grid
		{4, 4}	square grid
		{6, 3}	hexagonal grid
004 - 021	triangulation	$\{4, 4\}_1$	
		$\{4, 4\}_2$	tetrakis square tiling
		$\{6, 3\}_{1-13}$	
		$\{6, 3\}_{14}$	triakis triangular tiling
		$\{6, 3\}_{15}$	
		$\{6, 3\}_{16}$	bisected hexagonal tiling

cat. number	grid class	name and type	references/ notes
001 - 008	semiregular tiling	{3·6·3·6}	
		{3·3·3·3·6}	
		{3·3·3·4·4}	
		{3·3·4·3·4}	
		{3·4·6·4}	
		{4·8·8}	
		{3·12·12}	
		{4·6·12}	
009 - 024	semiregular dual	{3·6·3·6} + quasiregular rhombic tiling	union
		quasiregular rhombic tiling	3-fold symmetry, 6-fold symmetry
		{3·3·3·3·6} + floret pentagonal tiling	union
		floret pentagonal tiling	6-fold irregular pentagon tiling
		{3·3·3·4·4} + prismatic pentagonal tiling	union
		prismatic pentagonal tiling	3-fold irregular pentagon tiling
		{3·3·4·3·4} + cairo pentagonal tiling	union
		cairo pentagonal tiling	4-fold irregular pentagon tiling
		{3·4·6·4} + deltoidal trihexagonal tiling	union
		deltoidal trihexagonal tiling	{3, 6}, {6, 3} triangulation
		{4·8·8} + tetrakis square tiling	union
		tetrakis square tiling	{4, 4} triangulation
		{3·12·12} + triakis triangular tiling	union
		triakis triangular tiling	{3, 6}, quasiregular rhombic tiling
		{4·6·12} + bisected hexagonal tiling	union
		bisected hexagonal tiling	{3, 6}, {6, 3} triangulation
001 - 026	tiling variation	{3, 6} scalene triangles	*
		{3, 6} equilateral and isosceles triangles	*
		{3, 6} scalene and isosceles triangles	*
		{3, 6} equilateral, isosceles and scalene triangles	*
		{3·3·3·4·4} 4.147mm length increment squares	
		{3·3·3·4·4} 150% length increment squares	
		{3·3·3·4·4} x^y length increment squares	
		{3·3·3·4·4} rectangles and isosceles triangles	*
		{3·3·4·3·4} isosceles triangles	*
		{4, 4} rectangles and trapezoids	*
		{4, 4} rectangles and right triangles	*
		{6, 3} irregular hexagons	*
		{3·4·6·4} rectangles	*
		{4·6·12} rectangles and irregular hexagons	*
		3 square grids	0° offset, 46° offset, 80° offset
		3 square grids	0° offset, 63° offset, 77° offset
		{3·3·3·4·4} + {3·3·3·4·4} 4.147mm length increment	union
		{3·3·3·4·4} + {3·3·3·4·4} 150% length increment	union
		non-regular octogon {4·8·8} + non-regular octogon {4·8·8} 50% width decrement	union
		regular octogon {4·8·8} + regular octogon {4·8·8} 50% width decrement	union
		{3·12·12} + {3·12·12} 50% length decrement	union
		non-regular dodecagon {4·6·12} + non-regular dodecagon {4·6·12} 50% length decrement	union

* further reading [A.V. Shubnikov & V.A. Koptsik]

cat. number	grid class	name and type	references/ notes
001 - 026	tiling variation (cont.)	isometric grid to square grid	{3, 6} vertex displacement
		isometric grid to square grid	{3, 6} progressive skewing
		hexagon grid to non-regular hexagon grid	{6, 3} vertex displacement
		hexagon grid to non-regular hexagon grid	{6, 3} progressive skewing
		2 isometric grids to square grids	union
		2 hexagon grids to non-regular hexagon grids	union
001 - 124	demiregular tiling		$k = n$ transitivy classes
001 - 020		$k = 2$ uniform tilings	
021 - 059		$k = 3$ uniform tilings	
060 - 092		$k = 4$ uniform tilings	
093 - 107		$k = 5$ uniform tilings	
108 - 117		$k = 6$ uniform tilings	
118 - 124		$k = 7$ uniform tilings	
001 - 004	pentagonal tiling	dihedral pentagonal tiling	pentiling
		dürer pentagonal tiling	pentiling
		kepler pentagonal tiling	pentiling
		kepler–penrose pentagonal tiling	pentiling
001 - 042	quasiperiodic tiling	3-fold symmetry, 6-fold symmetry	quasiregular rhombic tiling
002 - 012		5-fold symmetry	unions
013 - 024		7-fold symmetry	unions
025 - 026		8-fold symmetry	ammann–beenker tilings
027 - 036		9-fold symmetry	unions
037 - 042		12-fold symmetry	
		socolar tiling	
		socolar tiling + socolar tiling inflation	union
		shield tiling	
		plate tiling	
		shield tiling + socolar tiling	union
		plate tiling + socolar tiling	union
001 - 013	non-periodic tiling	penrose rhombus tiling	
		penrose kite and dart tiling	
		3 penrose kite and dart tilings	0° offset, 36° offset, 72° offset
004 - 013		10-fold symmetry	spiky decagon tiling **

** further reading [A. Losev]

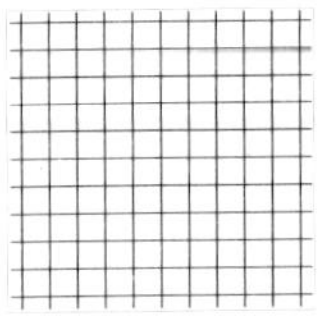
/gi-og

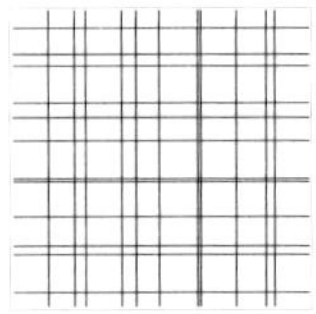
/gi-og/_union

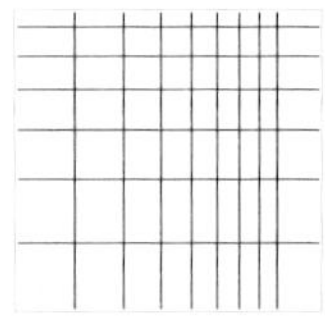
/gi-og/_log

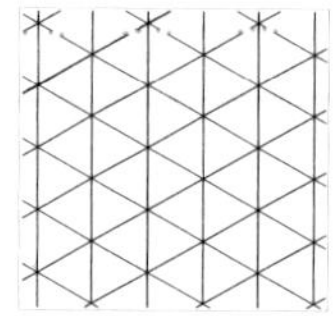
/gi-rt

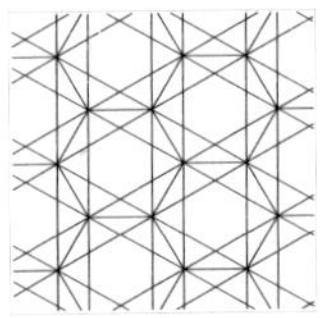
/gi-rt/_triangulate

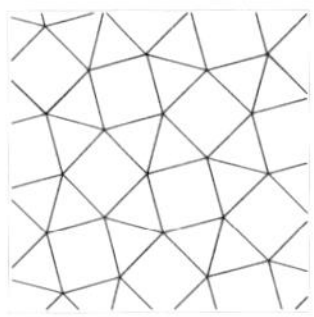
/gi-st

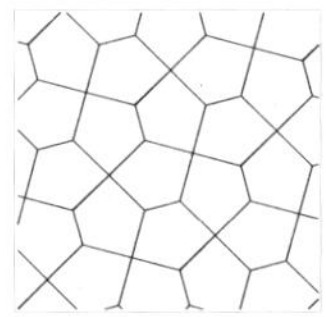
/gi-st/_duals

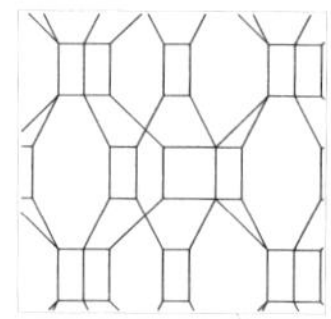
/gi-va

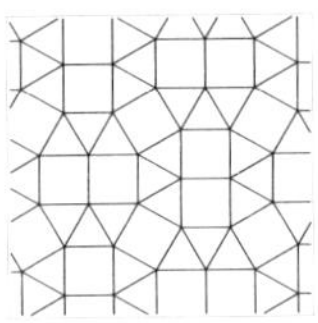
/gi-dt/_k2

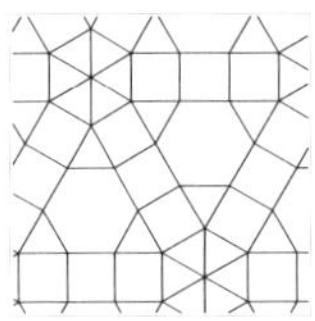
/gi-dt/_k3

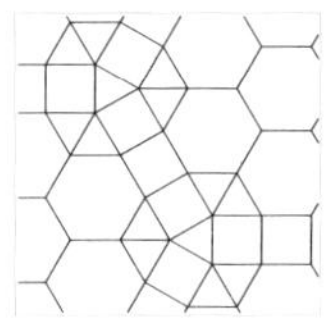
/gi-dt/_k4

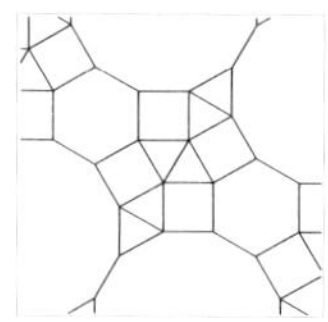
/gi-dt/_k5

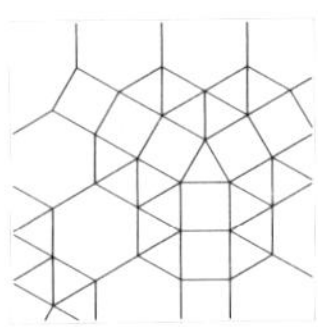
/gi-dt/_k6

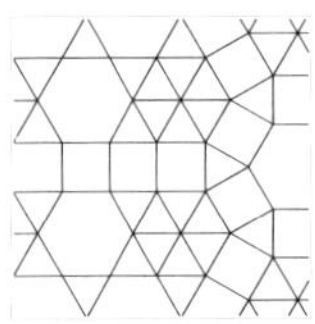
/gi-dt/_k7

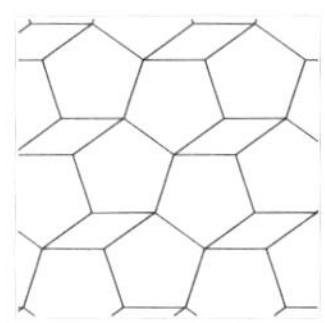
/gi-pt

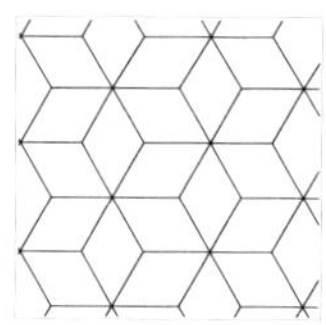
/gi-qt/_3fold

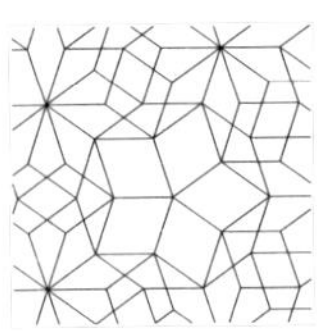
/gi-qt/_5fold

/gi-qt/_7fold

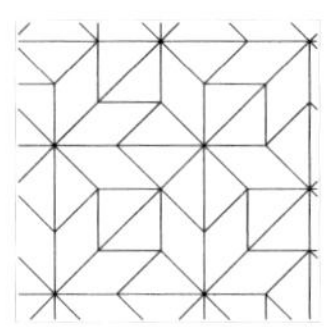
/gi-qt/_8fold

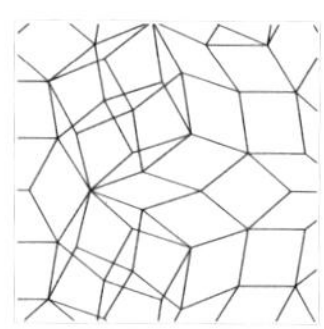
/gi-qt/_9fold

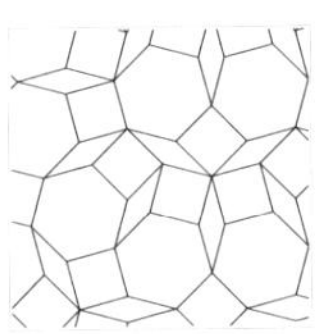
/gi-qt/_12fold

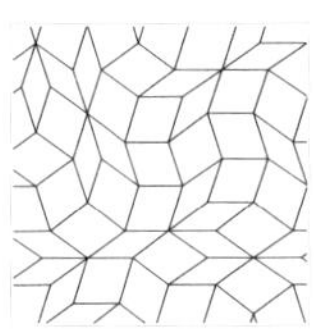
/gi-npt/_penrose

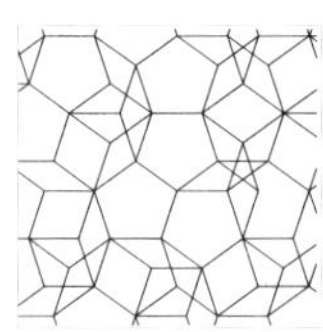
/gi-npt/_10fold

this set of software comes with the "carsten nicolai . grid index" publication.

title: grid index
author: carsten nicolai
published by Die Gestalten Verlag GmbH & Co KG

ISBN: 978-3-89955-241-6

hardcover/ 312 pages/ 185x230x38 mm
includes 1 CD containing 298 free editable .svg files and 298 .pdf files
First release: March 2009

_licence

this document allows the customer of "grid index" the right to install and use this software on 1 (one) computer platform. for further details see below.

_system requirements

MacOSX, Windows XP (or higher); Adobe Illustrator Legacy version 10 (or higher)

_legal notice

in the remainder of this document the term "software" refers to all or any digital material on this data disc.

_software product licence

this software is protected by copyright laws and international copyright treaties, as well as other intellectual property laws and treaties. the software is licensed, not sold.

_grant of licence

this document grants you, the user, the following rights:
a. installation and use:
you may install and use a single copy of the software.
you may use the software for every kind of commercial and/or non-commercial publication; electronic and/or print.

b. reproduction and distribution:
you cannot reproduce or distribute copies of the software.
you cannot make the software available for download.
copies of the software not be distributed for profit either on a standalone basis or included as part of your own product unless by prior permission of the author.

_description of other rights and limitations

- restrictions on alteration:
you may not rename, edit or create any derivative works from the software product, other than subsetting when embedding them in documents, unless you have permission from the author.

_limited warranty

carsten nicolai and the publisher warrant that the software product will operate under the system requirements specified above. no other warranty, express or implied, is made with respect to the software, including without limitation any implied warranty of merchantability, fitness for a particular purpose, non-infringement of third party rights and those arising from a course of dealing or usage of trade. no warranty is made that the use of the licenced materials will be error free or uninterrupted, that any errors or defects in the licenced materials will be corrected, or that the licenced materials' functionality will meet your requirements. the entire risk arising out of use or performance of the software is of the sole responsibility of the end-user.
in no event shall carsten nicolai and/or the publisher be liable for any damages whatsoever (including without limitation, damages for loss of business profits, business interruption, loss of business information or any other pecuniary loss) arising out of the use of or inability to use this product.

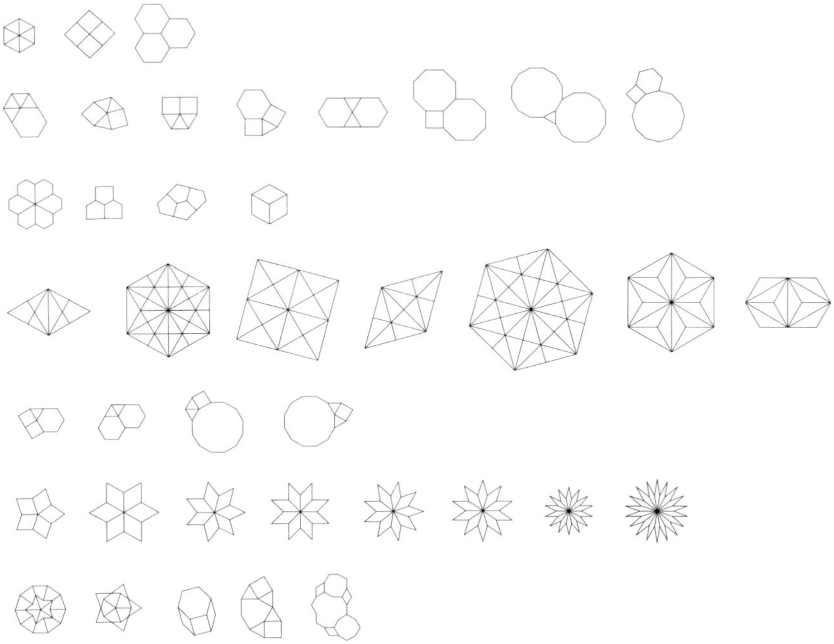

gi-og-001 orthogonal grid 100mm

gi-og-002 orthogonal grid 50mm

gi-og-003 orthogonal grid 25mm

gi-og-004 orthogonal grid 10mm

gi-og-005 orthogonal grid 5mm

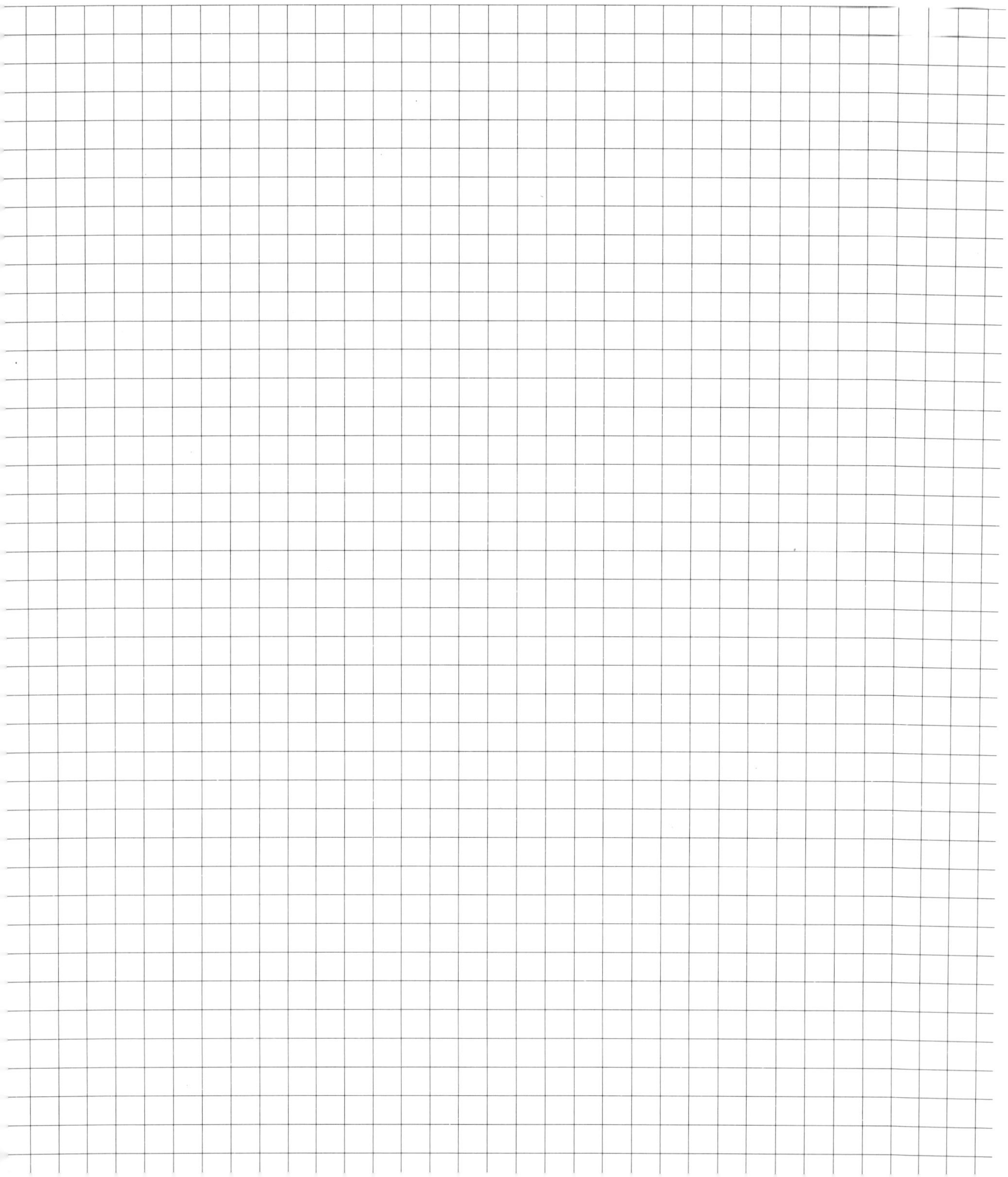

gi-og-006 orthogonal grid 2.5mm

gi-og-007 orthogonal grid 1mm

gi-og-008 orthogonal grid 1pt

gi-og-009 orthogonal grid 1 agate = 5.143pt = 1.814mm = 1/14 inch gi-og-010 orthogonal grid 1 pica = 12pt = 4.233mm = ⅙ inch

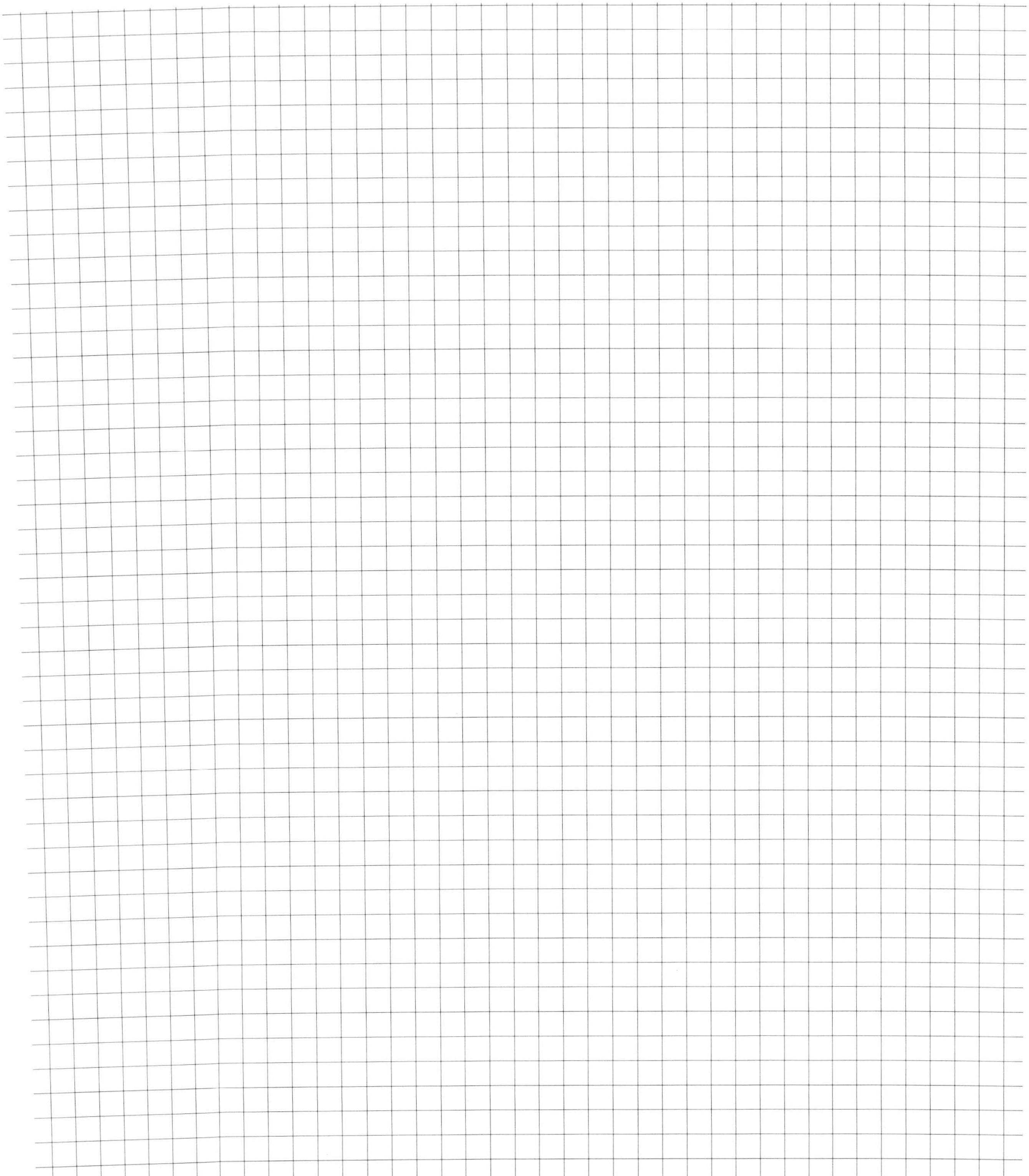

gi-og-011 orthogonal grid 1 cicéro = 12.788pt = 4.511mm
= ⅙ inch (france)

gi-og-013 orthogonal grid ½ inch

gi-og-014 orthogonal grid 1 inch

gi-og-015 orthogonal grid 2 inch

gi-og-017

orthogonal grid
union

100mm
4 inch

gi-og-018 orthogonal grid
union

50mm
2 inch

gi-og-019 orthogonal grid 25mm
union 1 inch

gi-og-020

orthogonal grid
union

10mm
½ inch

gi-og-021

orthogonal grid
union

5mm
¼ inch

gi-og-022 orthogonal grid 2.5mm
union 1 pica

gi-og-023

orthogonal grid
union

1 agate
1 pica
1 cicéro

gi-og-024 orthogonal grid 1mm
union 1pt

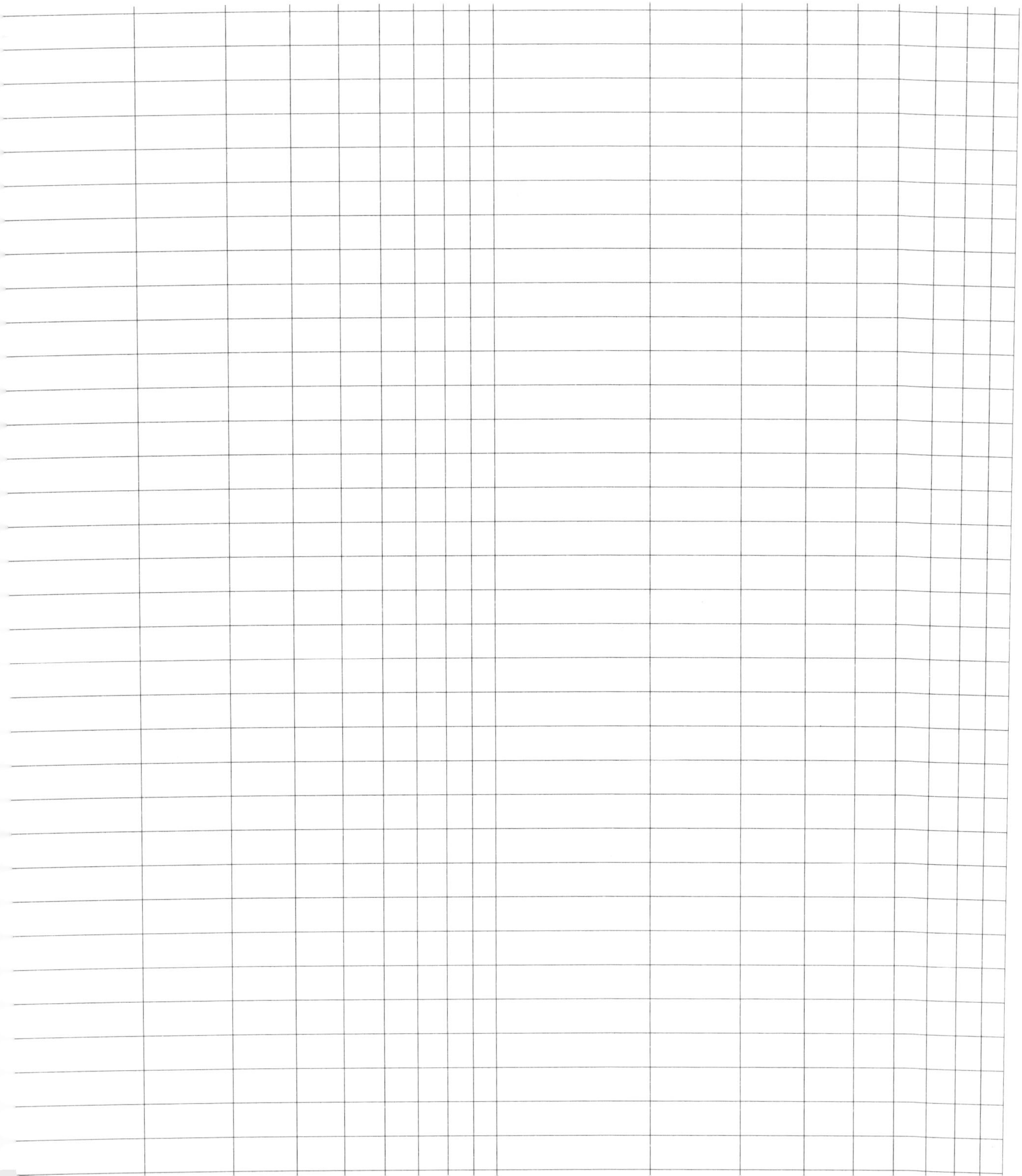

gi-og-027 logarithmic grid 4 cycle semi-log

gi-og-028 logarithmic grid 5 cycle semi-log

gi-og-029 logarithmic grid 6 cycle semi-log

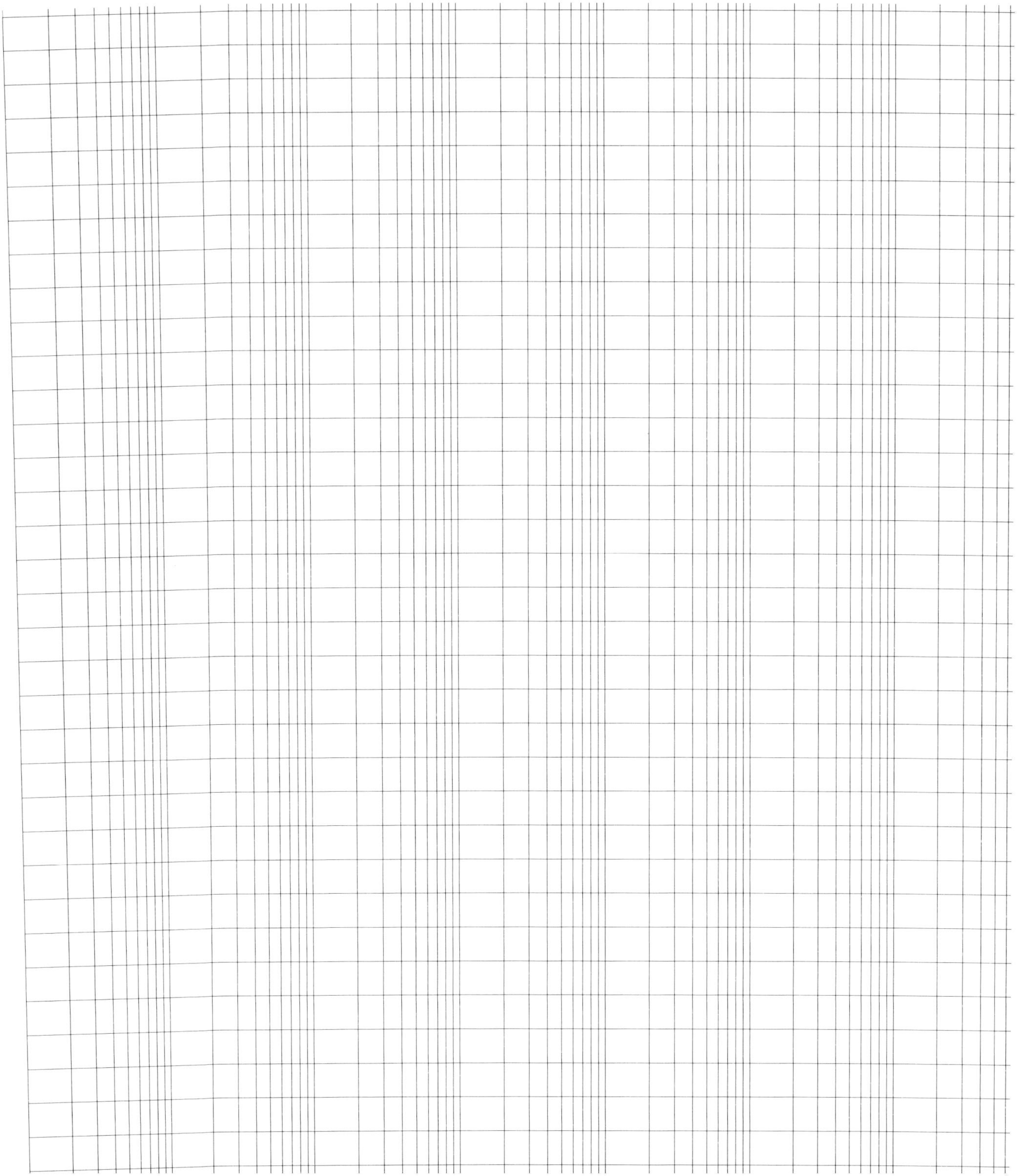

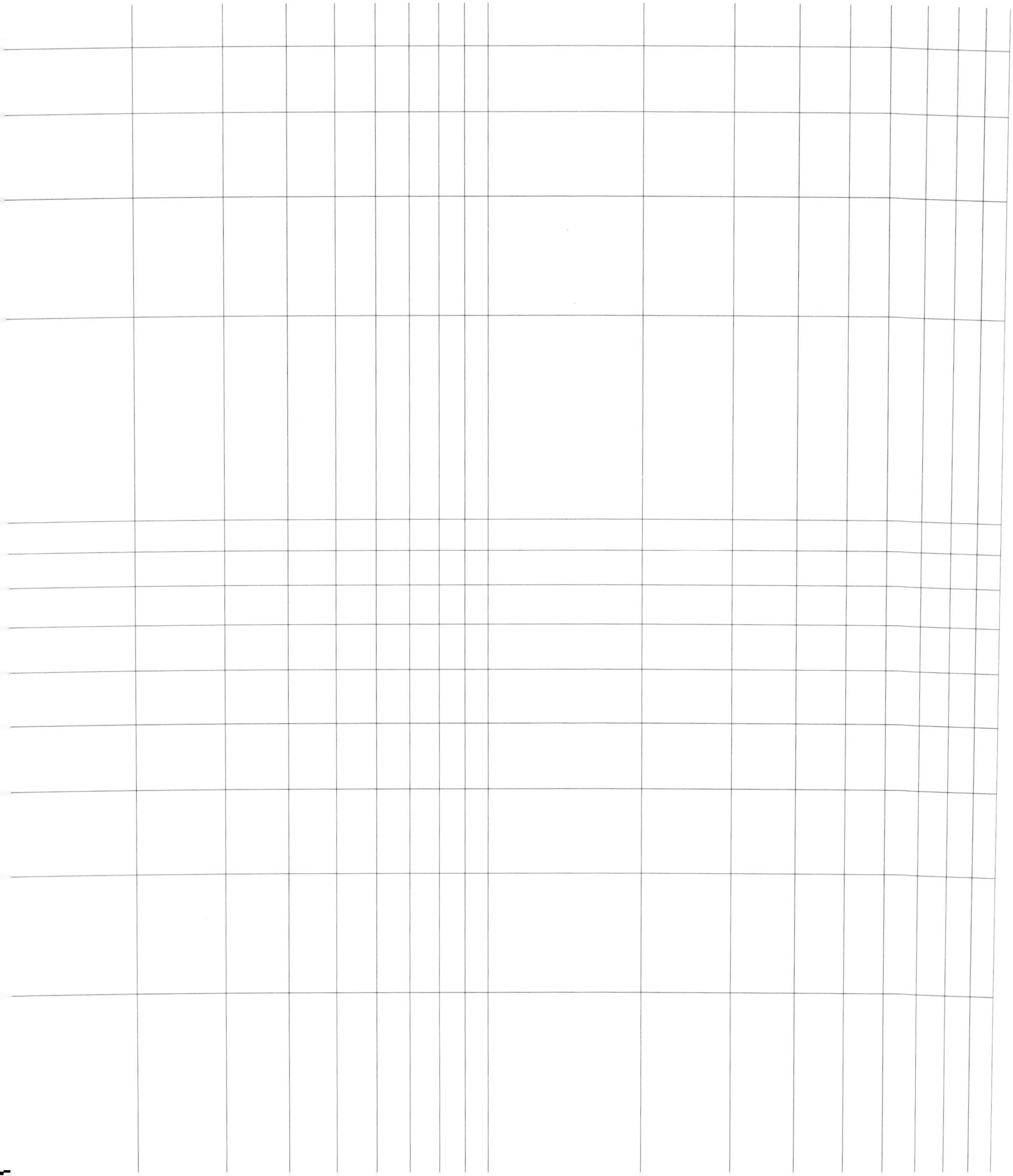

gi-og-033 logarithmic grid 4 cycle log-log

gi-og-034 logarithmic grid 5 cycle log-log

gi-og-036 logarithmic grid 7 cycle log-log

gi-rt-001 regular tiling {3, 6} isometric grid

gi-rt-004 regular tiling triangulation $\{4, 4\}_1$

regular tiling
triangulation

$\{4, 4\}_2$

tetrakis square tiling

gi-rt-003 regular tiling {6, 3} hexagonal grid

regular tiling
triangulation

{6, 3}$_1$

regular tiling
triangulation

$\{6, 3\}_2$

gi-rt-008 regular tiling triangulation $\{6, 3\}_3$

regular tiling
triangulation

$\{6, 3\}_4$

gi-rt-010 regular tiling triangulation $\{6, 3\}_5$

regular tiling
triangulation

$\{6, 3\}_6$

regular tiling
triangulation

$\{6, 3\}_7$

gi-rt-013 regular tiling triangulation $\{6, 3\}_8$

gi-rt-014 regular tiling {6, 3}$_9$
triangulation

gi-rt-015 regular tiling triangulation $\{6, 3\}_{10}$

gi-rt-016 regular tiling
triangulation

$\{6, 3\}_{11}$

gi-rt-017 regular tiling triangulation $\{6, 3\}_{12}$

regular tiling
triangulation

$\{6, 3\}_{13}$

regular tiling
triangulation

$\{6, 3\}_{14}$

triakis triangular tiling

gi-rt-020

regular tiling
triangulation

$\{6, 3\}_{15}$

regular tiling
triangulation

$\{6, 3\}_{16}$

bisected hexagonal tiling

semiregular tiling
dual tiling

{3·6·3·6}
quasiregular rhombic tiling

gi-st-010 dual tiling quasiregular rhombic tiling 3-fold symmetry
6-fold symmetry

semiregular tiling {3·3·3·3·6}

semiregular tiling | {3·3·3·3·6}
dual tiling | floret pentagonal tiling

gi-st-012 dual tiling floret pentagonal tiling 6-fold irregular pentagon tiling

gi-st-003 semiregular tiling {3·3·3·4·4}

gi-st-013

semiregular tiling
dual tiling

{3·3·3·4·4}
prismatic pentagonal tiling

gi-st-014 dual tiling prismatic pentagonal tiling 3-fold irregular pentagon tiling

semiregular tiling
dual tiling

{3·3·4·3·4}
cairo pentagonal tiling

gi-st-016 dual tiling cairo pentagonal tiling 4-fold irregular pentagon tiling

semiregular tiling
dual tiling

{3·4·6·4}
deltoidal trihexagonal tiling

gi-st-018 dual tiling deltoidal trihexagonal tiling {6, 3} with triangulation
{3, 6}

 semiregular tiling {4·8·8}

semiregular tiling {4·8·8}
dual tiling tetrakis square tiling

gi-st-020 dual tiling tetrakis square tiling {4, 4} with triangulation

semiregular tiling
dual tiling

{3·12·12}
triakis triangular tiling

dual tiling

triakis triangular tiling

quasiregular rhombic tiling
{3, 6}
{6, 3}

semiregular tiling {4·6·12}

semiregular tiling {4·6·12}
dual tiling bisected hexagonal tiling

 dual tiling bisected hexagonal tiling {6, 3} with triangulation
{3, 6}

gi-va-001 regular tiling variation {3, 6} scalene triangles [A.V. Shubnikov & V.A. Koptsik] isogon 30

gi-va-002

regular tiling
variation

{3, 6} equilateral and isosceles triangles

[A.V. Shubnikov & V.A. Koptsik]
isogon 31

{3, 6} scalene and isosceles triangles

[A.V. Shubnikov & V.A. Koptsik]
isogon 32

regular tiling
variation

{3, 6} equilateral, isosceles and scalene triangles

[A.V. Shubnikov & V.A. Koptsik]
isogon 12

semiregular tiling variation

{3·3·3·4·4} 4.147mm length increment squares

semiregular tiling
variation

{3·3·3·4·4} 150% length increment squares

semiregular tiling variation

{3·3·3·4·4} x^y length increment squares

semiregular tiling variation

{3·3·3·4·4} rectangles and isosceles triangles

[A.V. Shubnikov & V.A. Koptsik]
isogon 23

semiregular tiling
variation

{3·3·4·3·4} isosceles triangles

[A.V. Shubnikov & V.A. Koptsik]
isogon 19

regular tiling
variation

{4, 4} rectangles and trapezoids

[A.V. Shubnikov & V.A. Koptsik]
isogon 18

gi-va-011

regular tiling
variation

{4, 4} rectangles and right triangles

[A.V. Shubnikov & V.A. Koptsik]
isogon 20

regular tiling
variation

{6, 3} irregular hexagons

[A.V. Shubnikov & V.A. Koptsik]
isogon 03

semiregular tiling
variation

{3·4·6·4} rectangles

[A.V. Shubnikov & V.A. Koptsik]
isogon 02

semiregular tiling
variation

{4·6·12} rectangles and irregular hexagons

[A.V. Shubnikov & V.A. Koptsik]
isogon 01

gi-va-015	regular tiling	{4, 4}	0° offset
	variation	{4, 4}	46° offset
	union	{4, 4}	80° offset

gi-va-016

regular tiling	{4, 4}	0° offset
variation	{4, 4}	63° offset
union	{4, 4}	77° offset

semiregular tiling	{3·3·3·4·4}
variation	{3·3·3·4·4} 4.147mm length increment squares
union	

gi-va-018

semiregular tiling
variation
union

{3·3·3·4·4}
{3·3·3·4·4} 150% length increment squares

gi-va-019

semiregular tiling
variation
union

non-regular octogon {4·8·8}
non-regular octogon {4·8·8} 50% width decrement

gi-va-020

semiregular tiling
variation
union

regular octogon {4·8·8}
regular octogon {4·8·8} 50% width decrement

gi-va-021

semiregular tiling	{3·12·12}
variation	{3·12·12} 50% length decrement
union	

semiregular tiling
variation
union

non-regular dodecagon {4·6·12}
non-regular dodecagon {4·6·12} 50% length decrement

gi-va-024 regular tiling variation

hexagon grid to irregular pentagon grid

{6, 3} vertex displacement

regular tiling variation

hexagon grid to irregular hexagon grid

{6, 3} progressive skewing

gi-va-027

regular tiling
variation
union

isometric grid to square grid
isometric grid to square grid

{3, 6} vertex displacement
{3, 6} progressive skewing

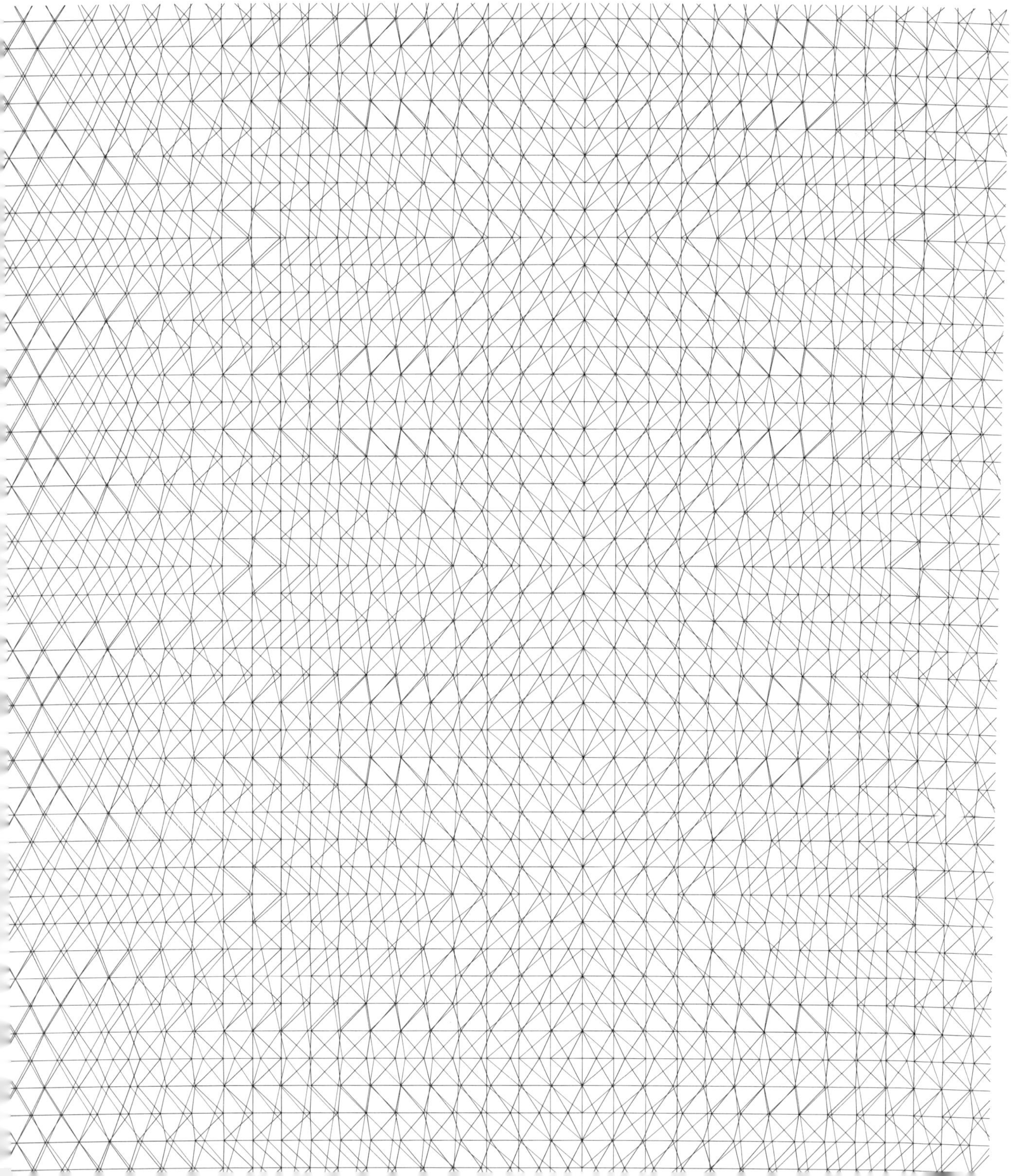

gi-va-028

regular tiling
variation
union

hexagon grid to irregular pentagon grid
hexagon grid to irregular hexagon grid

{6, 3} vertex displacement
{6, 3} progressive skewing

k=2 uniform tiling $\{3^3.4^2; 4^4\}_1$

k=2 uniform tiling $\{3^3{\cdot}4^2;\ 4^4\}_2$

k=2 uniform tiling $\{3^6; 3^3.4^2\}_2$

gi-dt-005 *k*=2 uniform tiling $\{3^3{\cdot}4^2; 3^2{\cdot}4{\cdot}3{\cdot}4\}_1$

 k=2 uniform tiling $\{3^3{\cdot}4^2; 3^2{\cdot}4{\cdot}3{\cdot}4\}_2$

k=2 uniform tiling $\{3^6; 3^2{\cdot}4{\cdot}3{\cdot}4\}$

gi-dt-008 *k*=2 uniform tiling $\{3^6; 3^2.6^2\}$

 k=2 uniform tiling $\{3^2{\cdot}6^2; 3{\cdot}6{\cdot}3{\cdot}6\}$

k=2 uniform tiling $\{3^4.6; 3^2.6^2\}$

k=2 uniform tiling $\{3^6; 3^4{\cdot}6\}_1$

k=2 uniform tiling $\{3^6; 3^4{\cdot}6\}_2$

k=2 uniform tiling {3·4²·6; 3·6·3·6}₁

gi-dt-016 *k*=2 uniform tiling $\{3{\cdot}4{\cdot}6{\cdot}4;\ 3^3{\cdot}4^2\}$

k=2 uniform tiling $\{3{\cdot}4{\cdot}6{\cdot}4;\ 3^2{\cdot}4{\cdot}3{\cdot}4\}$

k=2 uniform tiling $\{3{\cdot}4{\cdot}3{\cdot}12;\ 3{\cdot}12^2\}$

k=2 uniform tiling {3^6; 3^2·4·12}

k=2 uniform tiling {4·6·12; 3·4·6·4}

k=3 uniform tiling $\{3^6; 3^3{\cdot}4^2; 4^4\}_1$

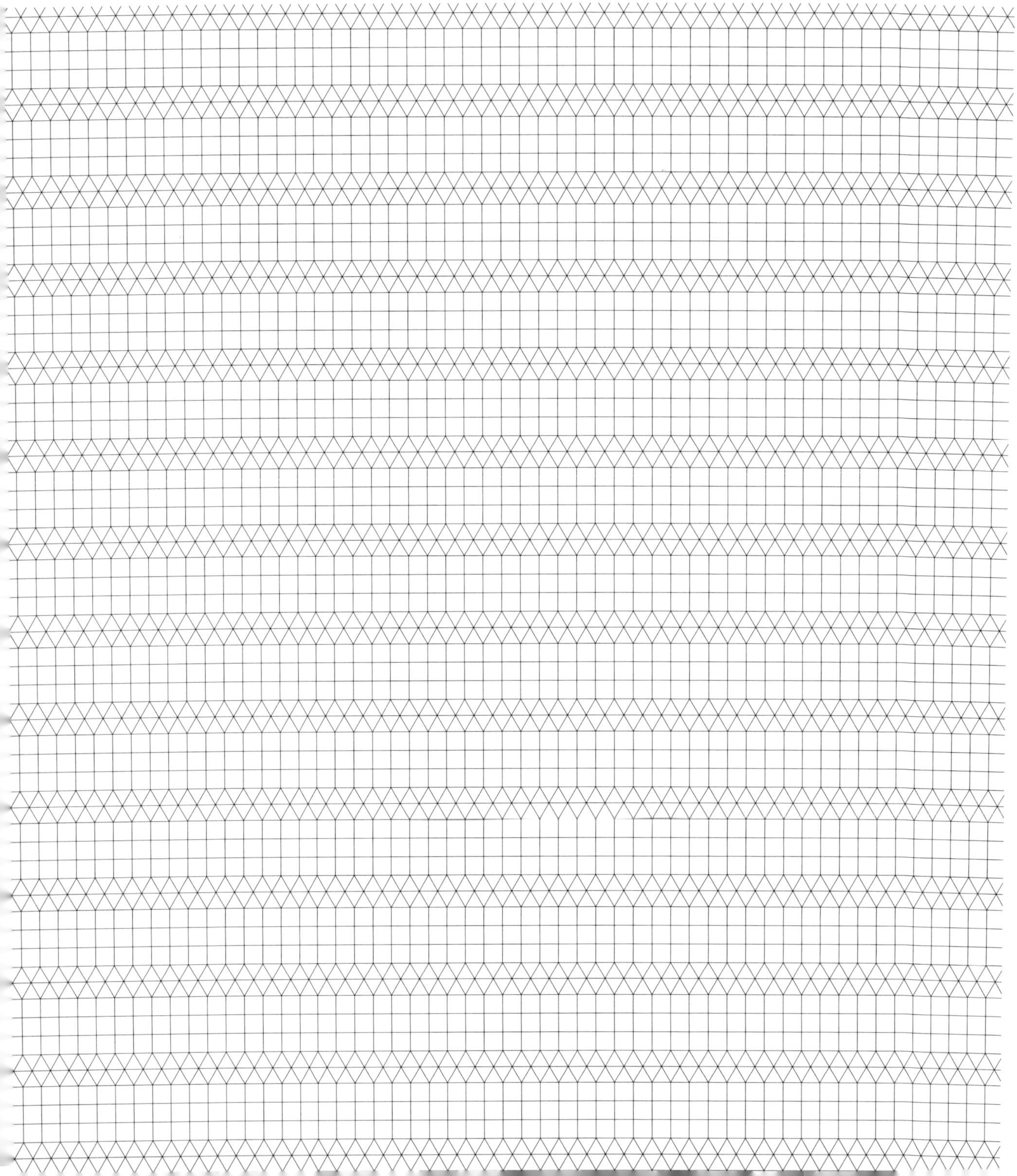

k=3 uniform tiling $\{3^6; 3^3.4^2; 4^4\}_2$

k=3 uniform tiling $\{3^6; 3^3{\cdot}4^2; 4^4\}_3$

k=3 uniform tiling $\{3^6; 3^3{\cdot}4^2; 4^4\}_4$

 k=3 uniform tiling {4^4; $3^3.4^2$; $3^2.4.3.4$}

k=3 uniform tiling $\{3^6; 3^2{\cdot}4{\cdot}3{\cdot}4; 3^3{\cdot}4^2\}$

k=3 uniform tiling $\{3^6; 3^2{\cdot}6^2; 3^4{\cdot}6\}_1$

k=3 uniform tiling $\{3^6; 3^2{\cdot}6^2; 3^4{\cdot}6\}_2$

 k=3 uniform tiling $\{3^6; 3^2{\cdot}6^2; 3^4{\cdot}6\}_3$

k=3 uniform tiling $\{3^6; 3^4{\cdot}6; 3{\cdot}6{\cdot}3{\cdot}6\}_3$

gi-dt-033 *k*=3 uniform tiling $\{6^3; 3^2 \cdot 6^2; 3 \cdot 6 \cdot 3 \cdot 6\}$

 k=3 uniform tiling $\{6^3; 3^2{\cdot}6^2; 3^4{\cdot}6\}$

k=3 uniform tiling $\{3^6; 3^2.6^2; 6^3\}$

k=3 uniform tiling $\{6^3; 3^2{\cdot}6^2; 3{\cdot}6{\cdot}3{\cdot}6\}$

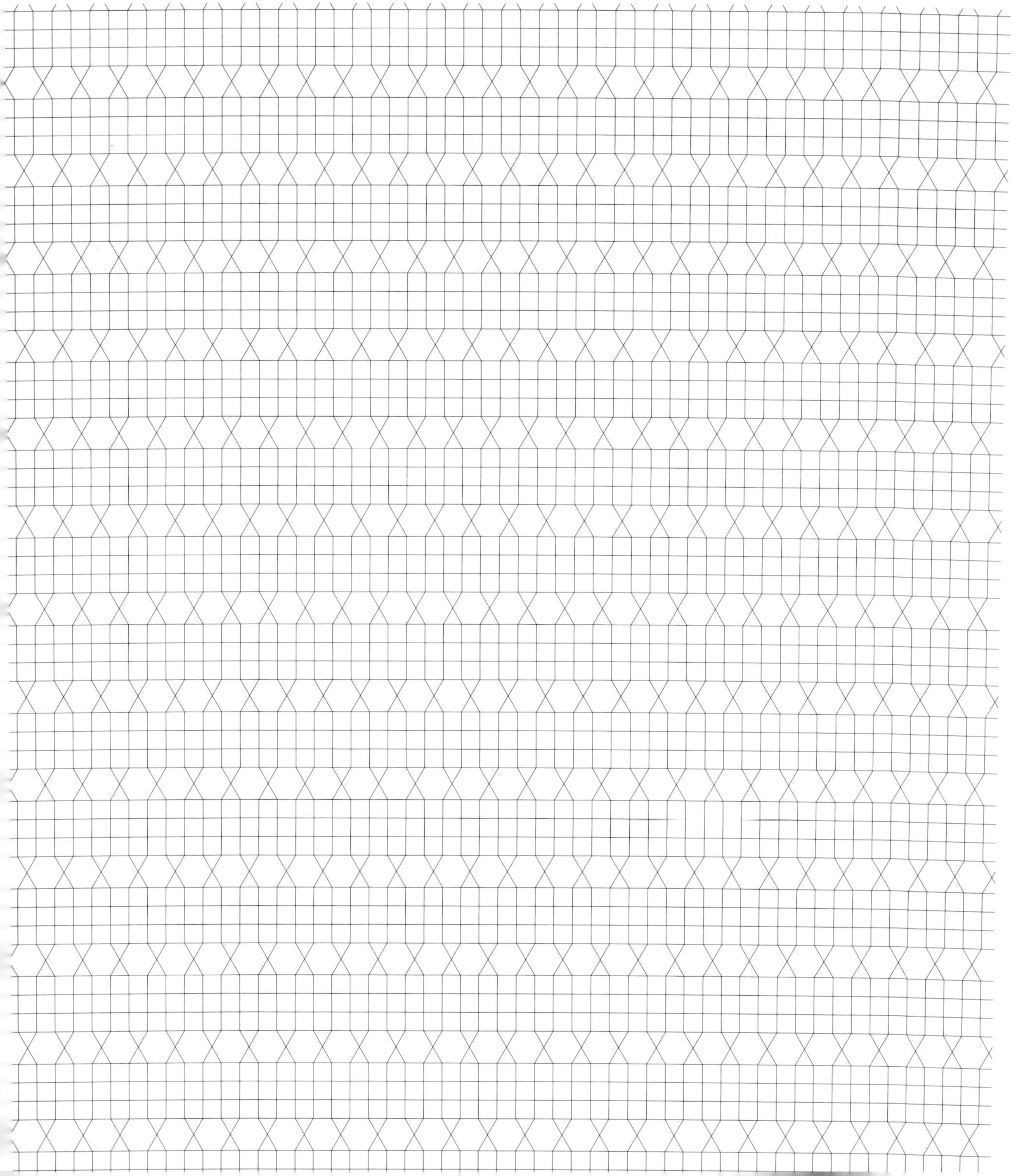

gi-dt-038 *k*=3 uniform tiling $\{4^4; 3{\cdot}4^2{\cdot}6; 3{\cdot}6{\cdot}3{\cdot}6\}_2$

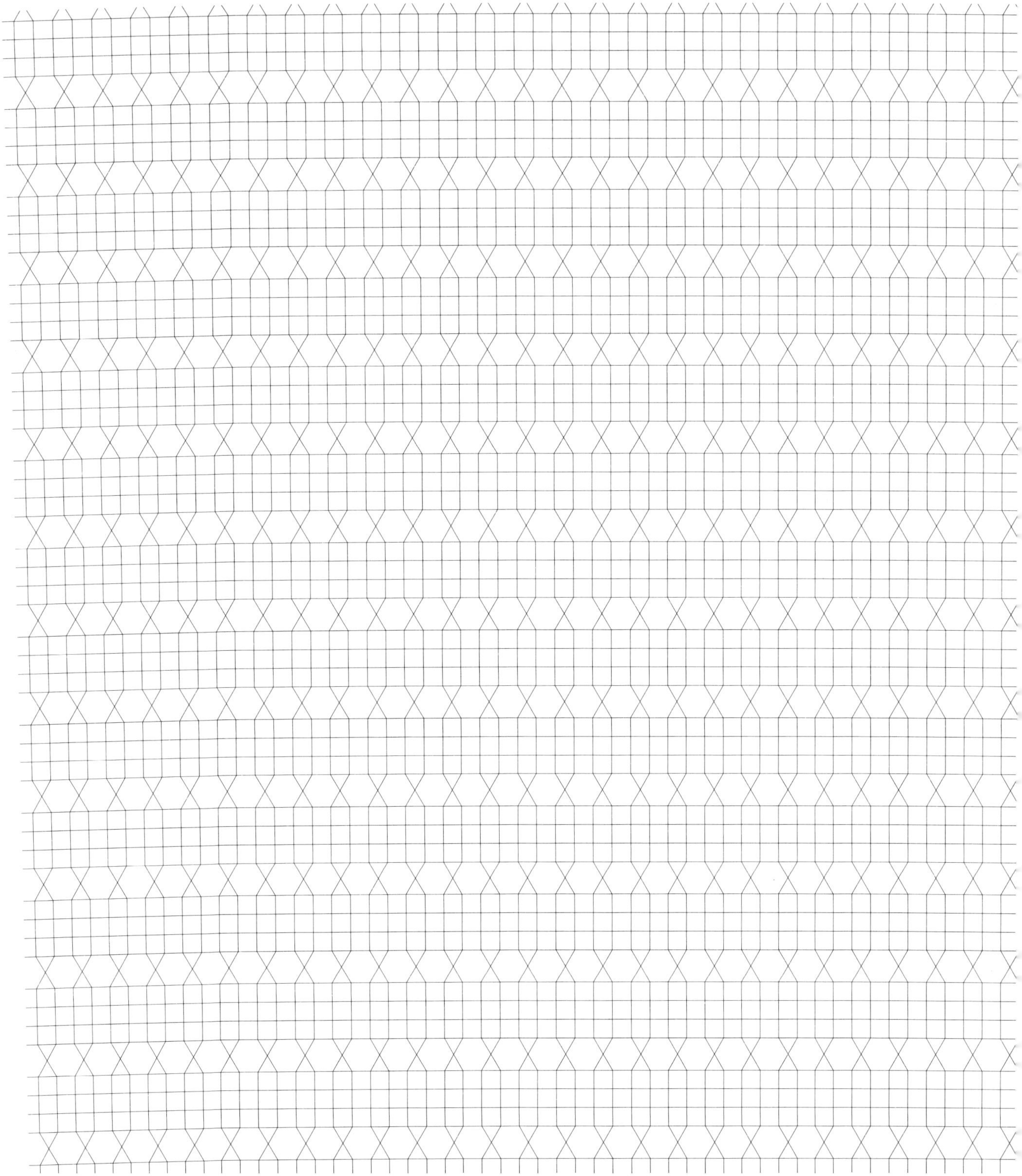

 k=3 uniform tiling $\{4^4; 3{\cdot}4^2{\cdot}6; 3{\cdot}6{\cdot}3{\cdot}6\}_3$

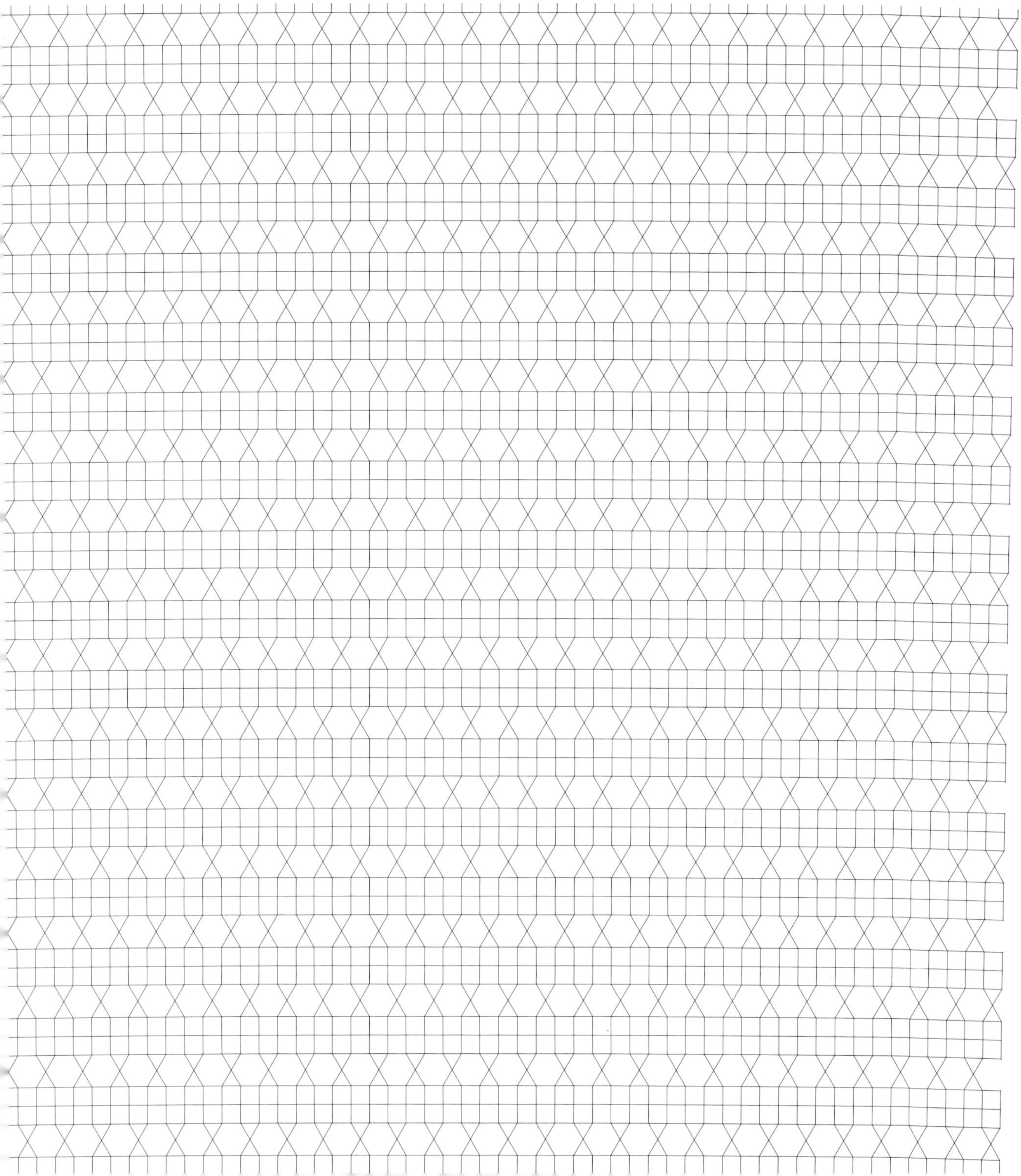

k=3 uniform tiling $\{4^4; 3{\cdot}4^2{\cdot}6; 3{\cdot}6{\cdot}3{\cdot}6\}_4$

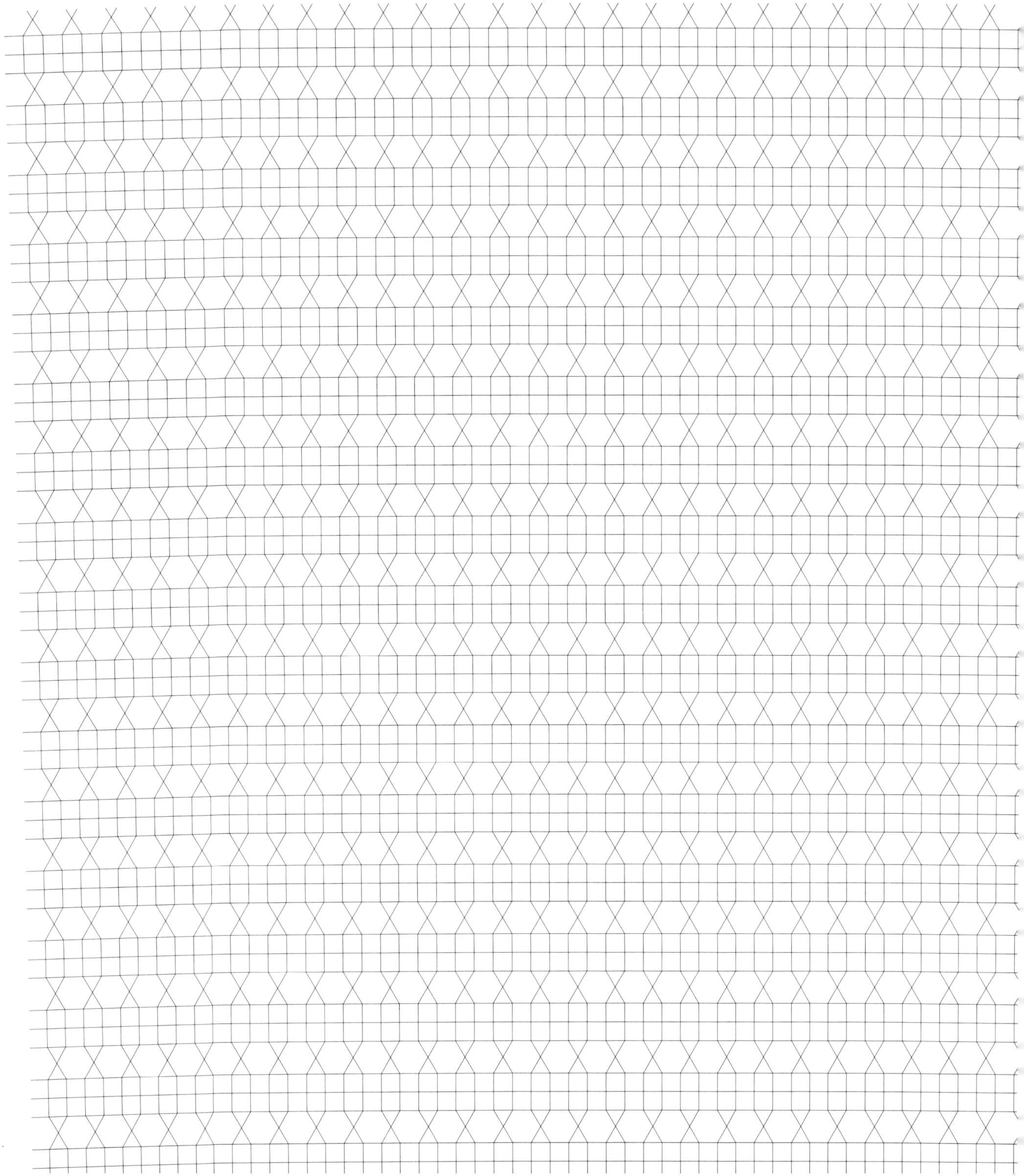

k=3 uniform tiling $\{3^3{\cdot}4^2;\ 3^2{\cdot}6^2;\ 3{\cdot}4^2{\cdot}6\}$

k=3 uniform tiling $\{3^2{\cdot}6^2;\ 3{\cdot}6{\cdot}3{\cdot}6;\ 3{\cdot}4^2{\cdot}6\}_1$

k=3 uniform tiling $\{3^2{\cdot}6^2;\ 3{\cdot}6{\cdot}3{\cdot}6;\ 3{\cdot}4^2{\cdot}6\}_2$

 k=3 uniform tiling $\{3^3{\cdot}4^2; 3^4{\cdot}6; 3{\cdot}4^2{\cdot}6\}$

k=3 uniform tiling $\{4^4; 3{\cdot}4^2{\cdot}6; 3{\cdot}4{\cdot}6{\cdot}4\}$

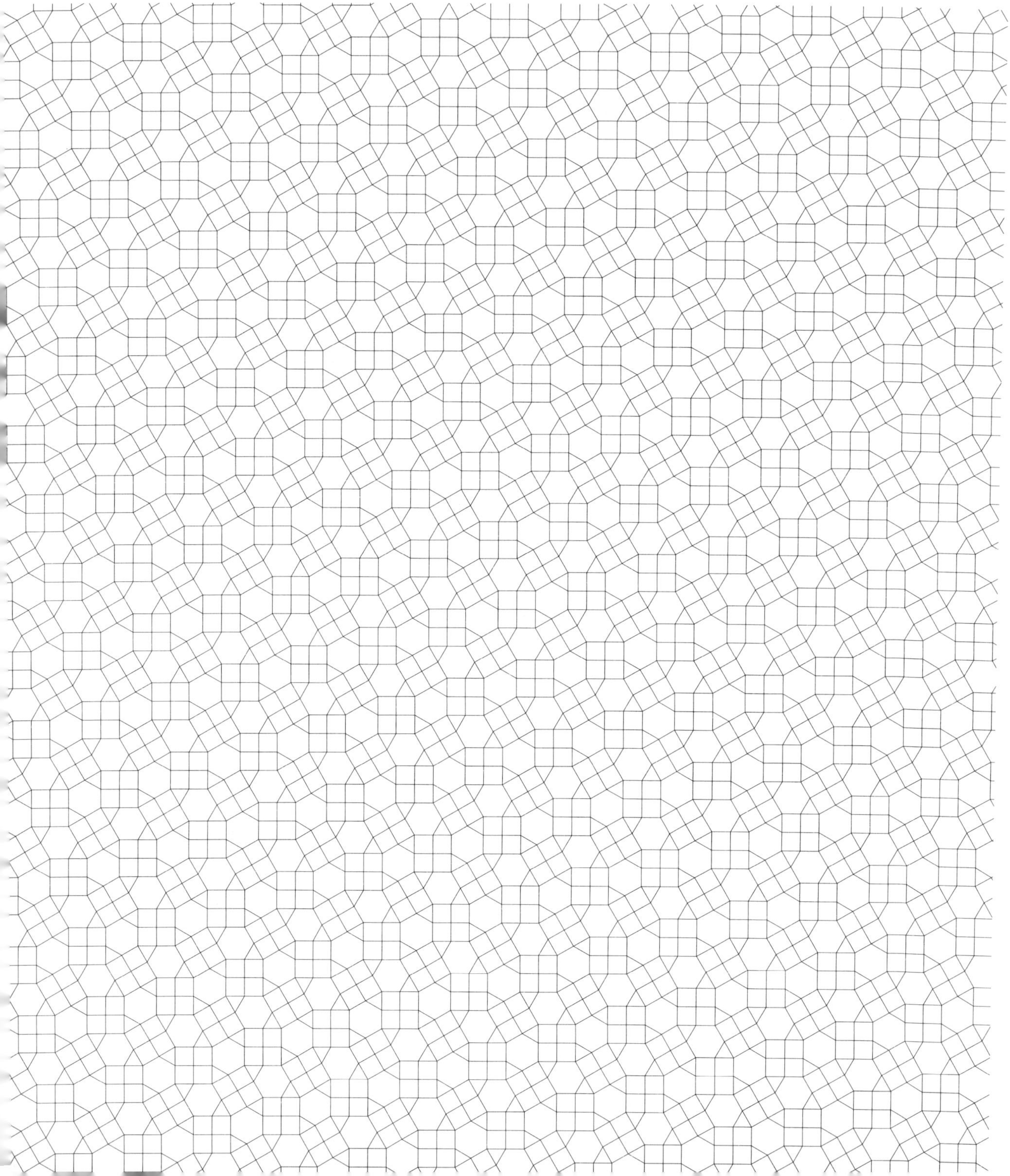

k=3 uniform tiling $\{3^6; 3^2{\cdot}4{\cdot}3{\cdot}4; 3{\cdot}4^2{\cdot}6\}$

k=3 uniform tiling {$3^2{\cdot}4{\cdot}3{\cdot}4$; $3{\cdot}4^2{\cdot}6$; $3{\cdot}4{\cdot}6{\cdot}4$}

k=3 uniform tiling $\{3^6; 3^2{\cdot}4{\cdot}3{\cdot}4; 3{\cdot}4{\cdot}6{\cdot}4\}_1$

k=3 uniform tiling {3^6; $3^3{\cdot}4^2$; $3{\cdot}4{\cdot}6{\cdot}4$}

k=3 uniform tiling $\{3^6; 3^2{\cdot}4{\cdot}3{\cdot}4; 3{\cdot}4{\cdot}6{\cdot}4\}_2$

k=3 uniform tiling

$\{3^3{\cdot}4^2;\ 3^2{\cdot}4{\cdot}3{\cdot}4;\ 3^4{\cdot}6\}$

 k=3 uniform tiling {3·12²; 3²·4·12; 3·4·3·12}

k=3 uniform tiling $\{3^6; 3^3{\cdot}4^2; 3^2{\cdot}4{\cdot}12\}$

k=3 uniform tiling $\{3^6; 3^2{\cdot}4{\cdot}3{\cdot}4; 3^2{\cdot}4{\cdot}12\}$

k=3 uniform tiling {3·6·3·6; 3^2·4·12; 4·6·12}

k=3 uniform tiling {3^6; 3^2·4·12; 3·4·6·4}

k=3 uniform tiling {3·12²; 3²·4·12; 3·4·6·4}

k=3 uniform tiling {3·12²; 3·4·3·12; 3·4·6·4}

k=3 uniform tiling $\{3^3{\cdot}4^2; 3^2{\cdot}4{\cdot}12; 3{\cdot}4{\cdot}6{\cdot}4\}$

k=4 uniform tiling $\{3^6; 4^4; 3^3{\cdot}4^2; 3^2{\cdot}4{\cdot}3{\cdot}4\}$

k=4 uniform tiling

$\{3^6; 3^4.6; 3^2.6^2; 3.6.3.6\}$

k=4 uniform tiling $\{3^6; 3^2{\cdot}6^2; 3^4{\cdot}6; 6^3\}_1$

k=4 uniform tiling $\{3^6; 3^2{\cdot}6^2; 3^4{\cdot}6; 6^3\}_2$

k=4 uniform tiling {6^3; $3^2 \cdot 6^2$; $3^4 \cdot 6$; $3 \cdot 6 \cdot 3 \cdot 6$}

 k=4 uniform tiling $\{3^6; 3^2{\cdot}6^2; 3^4{\cdot}6; 6^3\}_3$

k=4 uniform tiling $\{3^6; 3^2{\cdot}6^2; 3^4{\cdot}6; 6^3\}_4$

 k=4 uniform tiling {6^3; $3^2{\cdot}6^2$; $3{\cdot}6{\cdot}3{\cdot}6$; $3^4{\cdot}6$}

k=4 uniform tiling $\{4^4; 3^2{\cdot}6^2; 3{\cdot}6{\cdot}3{\cdot}6; 3{\cdot}4^2{\cdot}6\}_1$

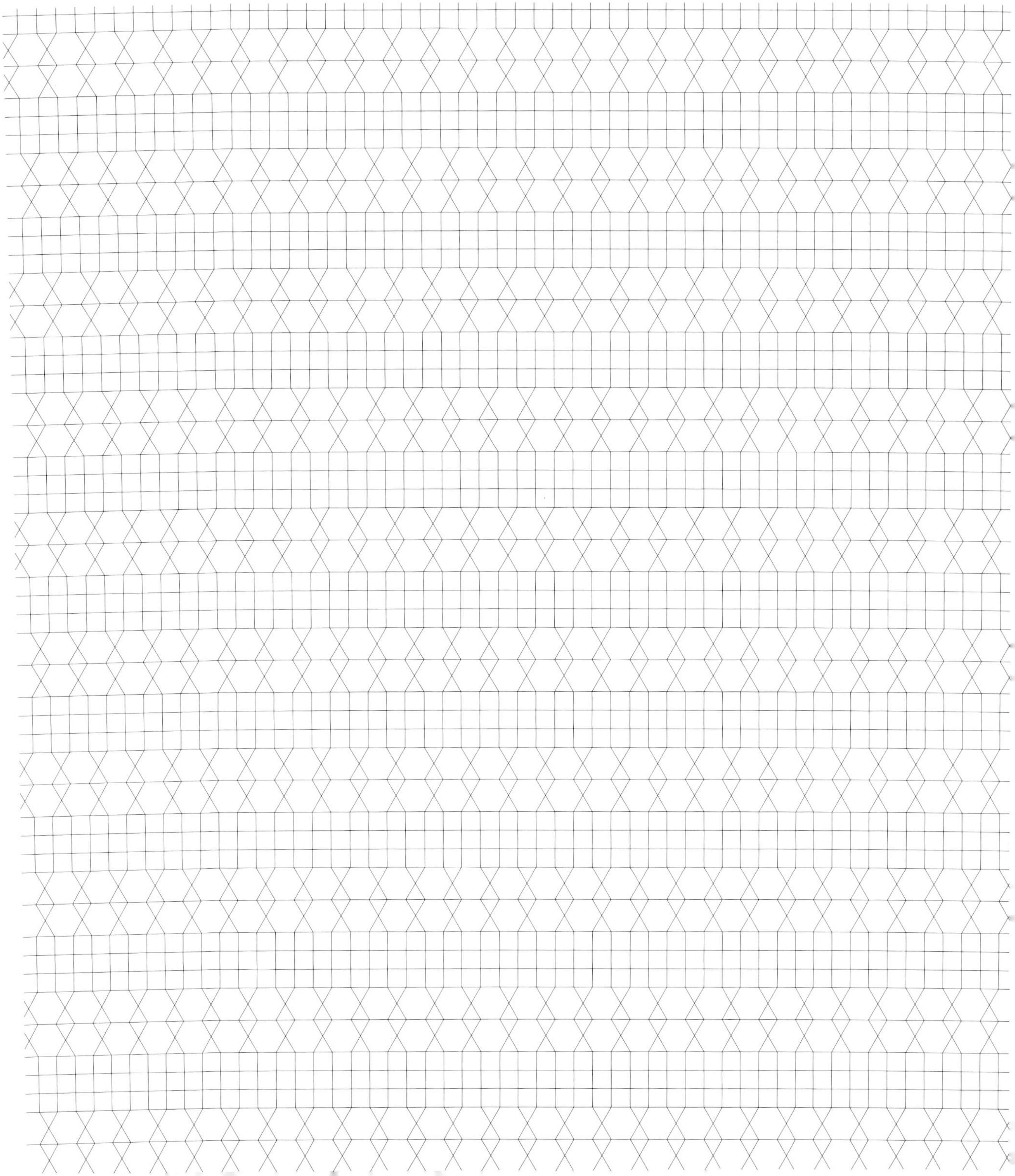

k=4 uniform tiling $\{4^4; 3^2{\cdot}6^2; 3{\cdot}6{\cdot}3{\cdot}6; 3{\cdot}4^2{\cdot}6\}_2$

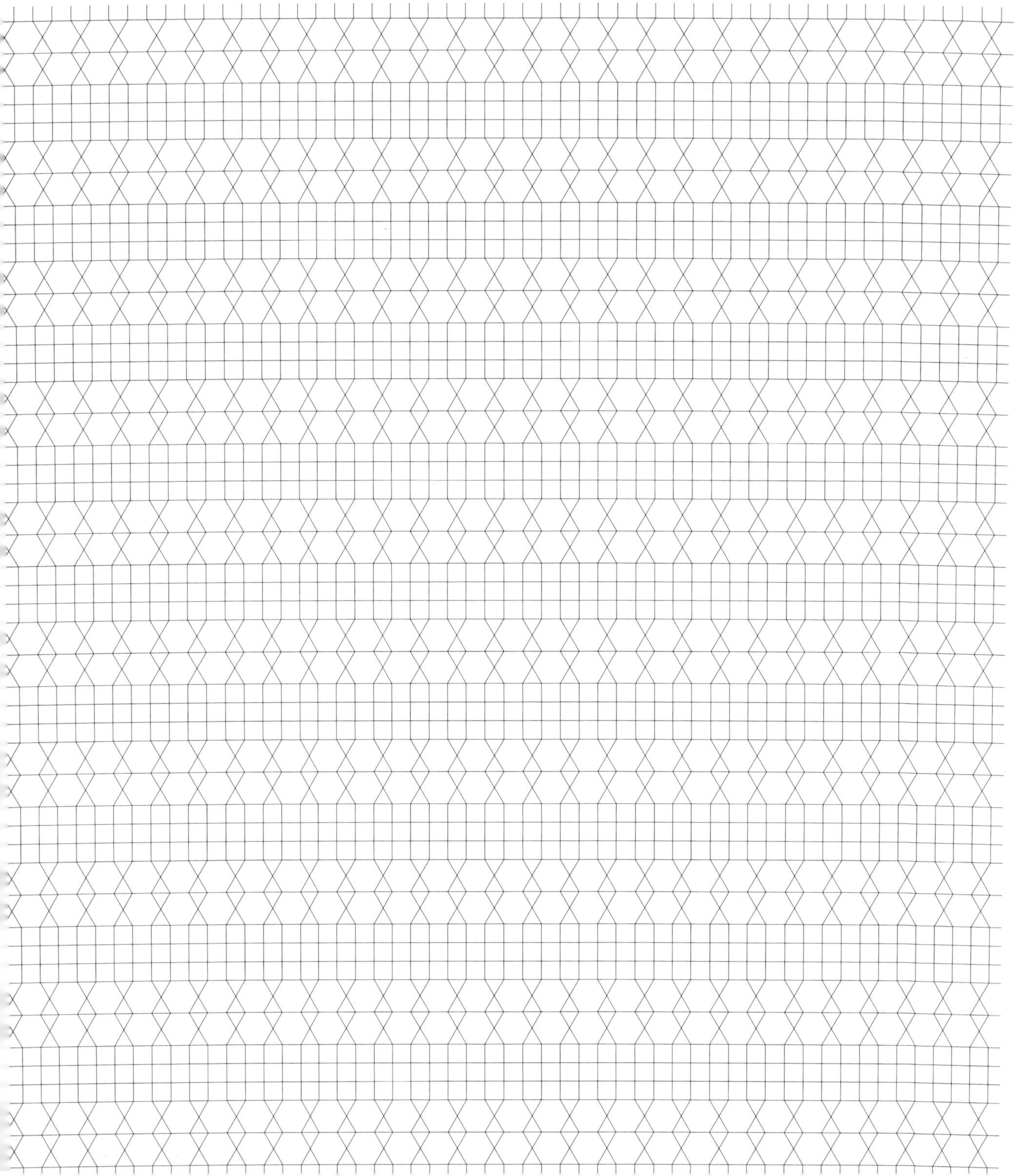

k=4 uniform tiling $\{4^4; 3^2{\cdot}6^2; 3{\cdot}6{\cdot}3{\cdot}6; 3{\cdot}4^2{\cdot}6\}_3$

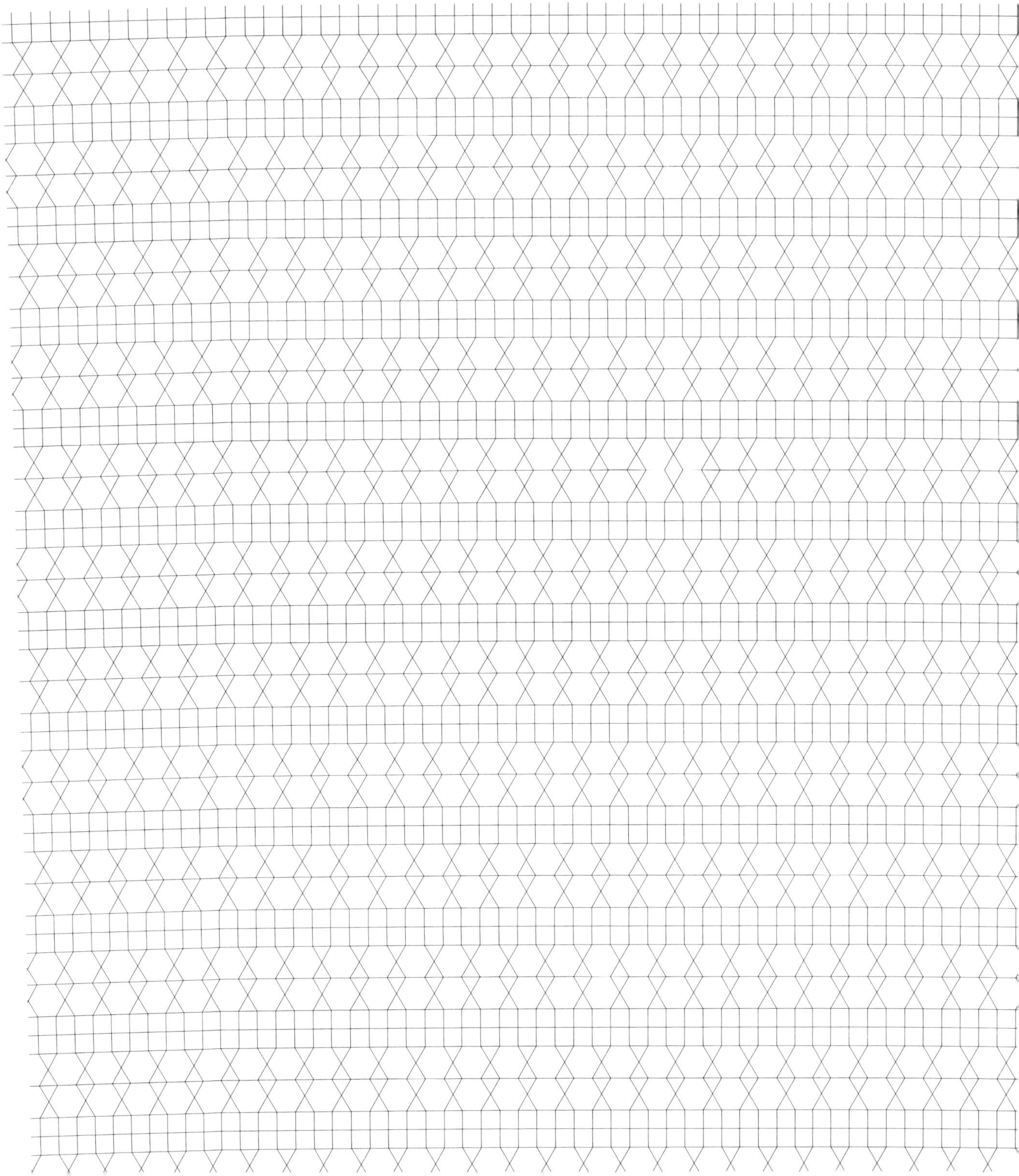

k=4 uniform tiling $\{4^4; 3^2{\cdot}6^2; 3{\cdot}6{\cdot}3{\cdot}6; 3{\cdot}4^2{\cdot}6\}_4$

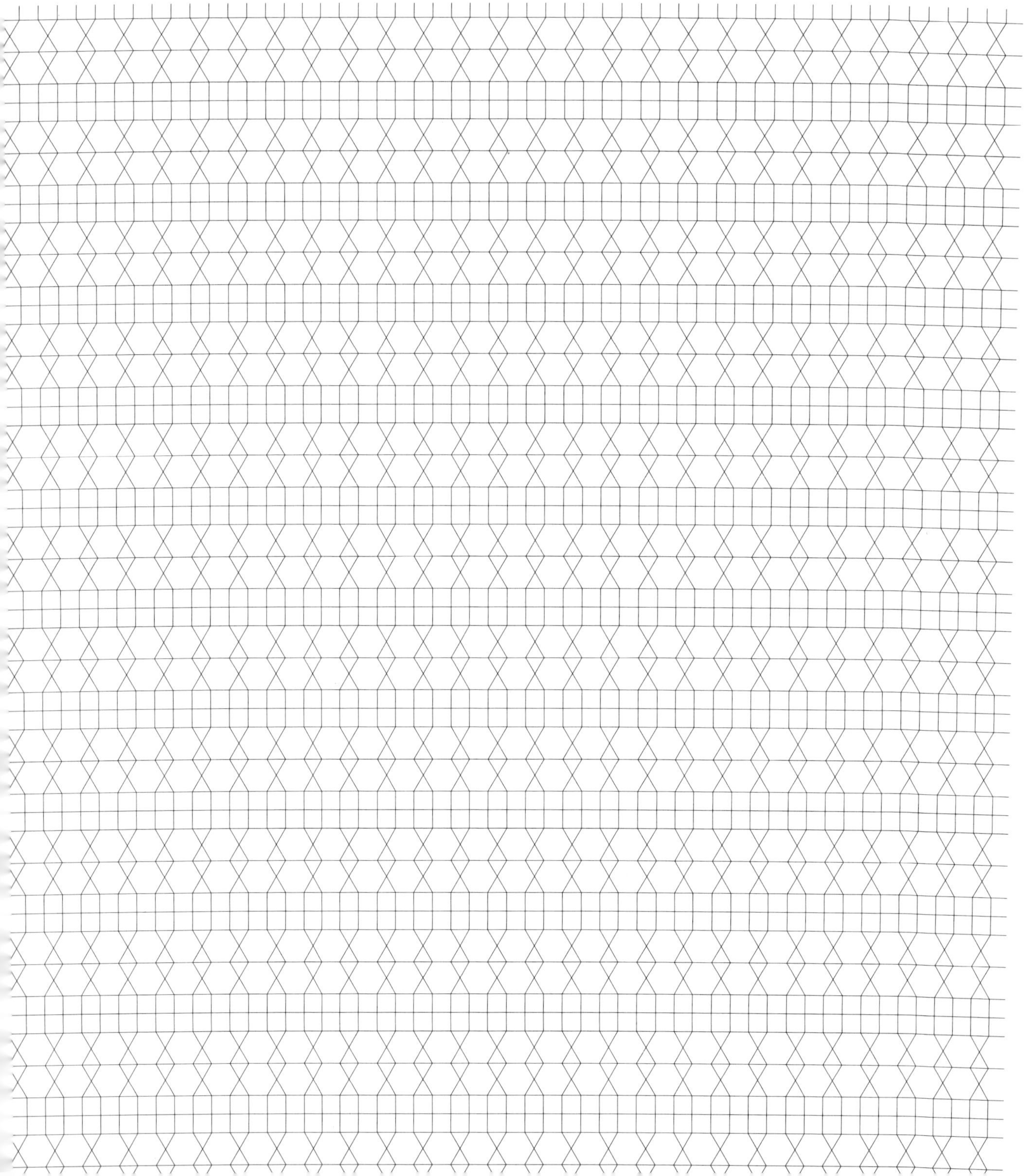

k=4 uniform tiling $\{3^6; 3^4{\cdot}6; 3{\cdot}6{\cdot}3{\cdot}6; 3{\cdot}4^2{\cdot}6\}_1$

 k=4 uniform tiling $\{3^6; 3^4.6; 3.6.3.6; 3.4^2.6\}_2$

k=4 uniform tiling

$\{3^6; 3^3{\cdot}4^2; 3{\cdot}6{\cdot}3{\cdot}6; 3{\cdot}4^2{\cdot}6\}$

gi-dt-075 k=4 uniform tiling $\{3^6; 3^3 \cdot 4^2; 3^4 \cdot 6; 3 \cdot 4^2 \cdot 6\}_1$

k=4 uniform tiling $\{3^6; 3^3{\cdot}4^2; 3^4{\cdot}6; 3{\cdot}4^2{\cdot}6\}_2$

k=4 uniform tiling {6^3; $3^2{\cdot}6^2$; $3^3{\cdot}4^2$; $3{\cdot}4^2{\cdot}6$}$_1$

k=4 uniform tiling $\{6^3; 3^2{\cdot}6^2; 3^3{\cdot}4^2; 3{\cdot}4^2{\cdot}6\}_2$

gi-dt-079 *k*=4 uniform tiling $\{3^6; 3^2{\cdot}4{\cdot}3{\cdot}4; 3{\cdot}4^2{\cdot}6; 3{\cdot}4{\cdot}6{\cdot}4\}$

k=4 uniform tiling {$3^2{\cdot}6^2$; $3^2{\cdot}4{\cdot}3{\cdot}4$; $3{\cdot}4^2{\cdot}6$; $3{\cdot}4{\cdot}6{\cdot}4$}

k=4 uniform tiling $\{3^6; 3^3{\cdot}4^2; 3^2{\cdot}4{\cdot}3{\cdot}4; 3{\cdot}4{\cdot}6{\cdot}4\}_1$

 k=4 uniform tiling $\{3^6; 3^3{\cdot}4^2; 3^2{\cdot}4{\cdot}3{\cdot}4; 3{\cdot}4{\cdot}6{\cdot}4\}_2$

k=4 uniform tiling {6^3; $3^2{\cdot}6^2$; $3^2{\cdot}4{\cdot}3{\cdot}4$; $3{\cdot}4^2{\cdot}6$}

 k=4 uniform tiling {3^6; $3^2{\cdot}4{\cdot}3{\cdot}4$; $3{\cdot}12^2$; $3^2{\cdot}4{\cdot}12$}

k=4 uniform tiling $\{3^6; 3^2{\cdot}4{\cdot}3{\cdot}4; 3{\cdot}12^2; 3{\cdot}4{\cdot}3{\cdot}12\}$

k=4 uniform tiling $\{3^3 \cdot 4^2; 3 \cdot 12^2; 3^2 \cdot 4 \cdot 12; 3 \cdot 4 \cdot 3 \cdot 12\}_1$

k=4 uniform tiling $\{3^3{\cdot}4^2; 3{\cdot}12^2; 3^2{\cdot}4{\cdot}12; 3{\cdot}4{\cdot}3{\cdot}12\}_2$

k=4 uniform tiling {4^4; $3^3{\cdot}4^2$; $3^2{\cdot}4{\cdot}12$; $3{\cdot}4{\cdot}3{\cdot}12$}

k=4 uniform tiling {3^6; $3^3{\cdot}4^2$; $3^2{\cdot}4{\cdot}3{\cdot}4$; $3^2{\cdot}4{\cdot}12$}

k=4 uniform tiling $\{3^3{\cdot}4^2; 3^2{\cdot}6^2; 3{\cdot}4^2{\cdot}6; 4{\cdot}6{\cdot}12\}$

k=4 uniform tiling $\{3^2{\cdot}4{\cdot}3{\cdot}4;\ 3^2{\cdot}6^2;\ 3{\cdot}4^2{\cdot}6;\ 4{\cdot}6{\cdot}12\}$

k=4 uniform tiling {3·4·6·4; 3^2·4·12; 3·4·3·12; 4·6·12}

k=5 uniform tiling

{3^6; 6^3; $3^2{\cdot}6^2$; $3{\cdot}6{\cdot}3{\cdot}6$; $3^4{\cdot}6$}

k=5 uniform tiling $\{3^6; 4^4; 3{\cdot}6{\cdot}3{\cdot}6; 3^4{\cdot}6; 3{\cdot}4^2{\cdot}6\}_1$

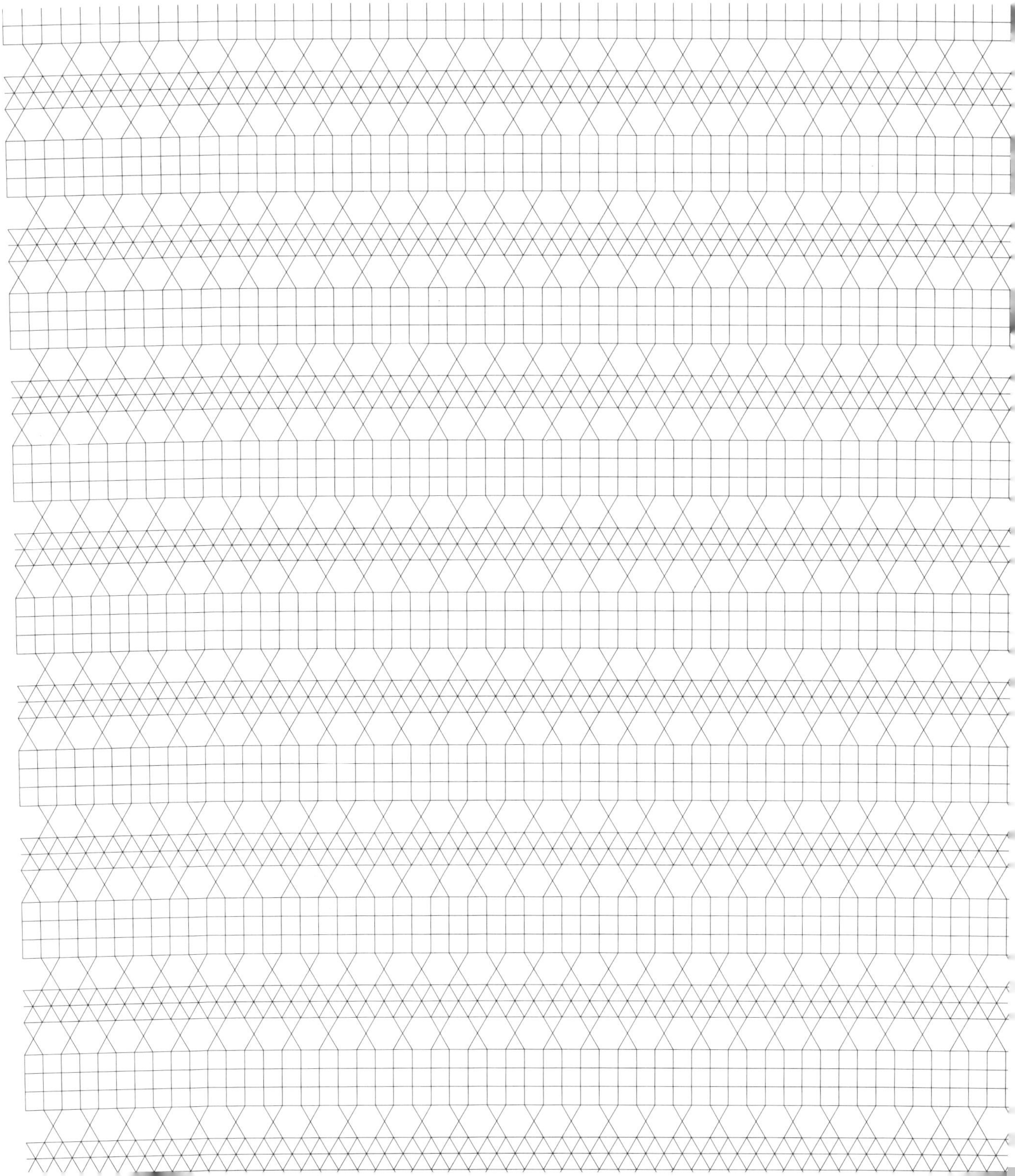

gi-dt-095 *k*=5 uniform tiling $\{3^6; 4^4; 3\cdot6\cdot3\cdot6; 3^4\cdot6; 3\cdot4^2\cdot6\}_2$

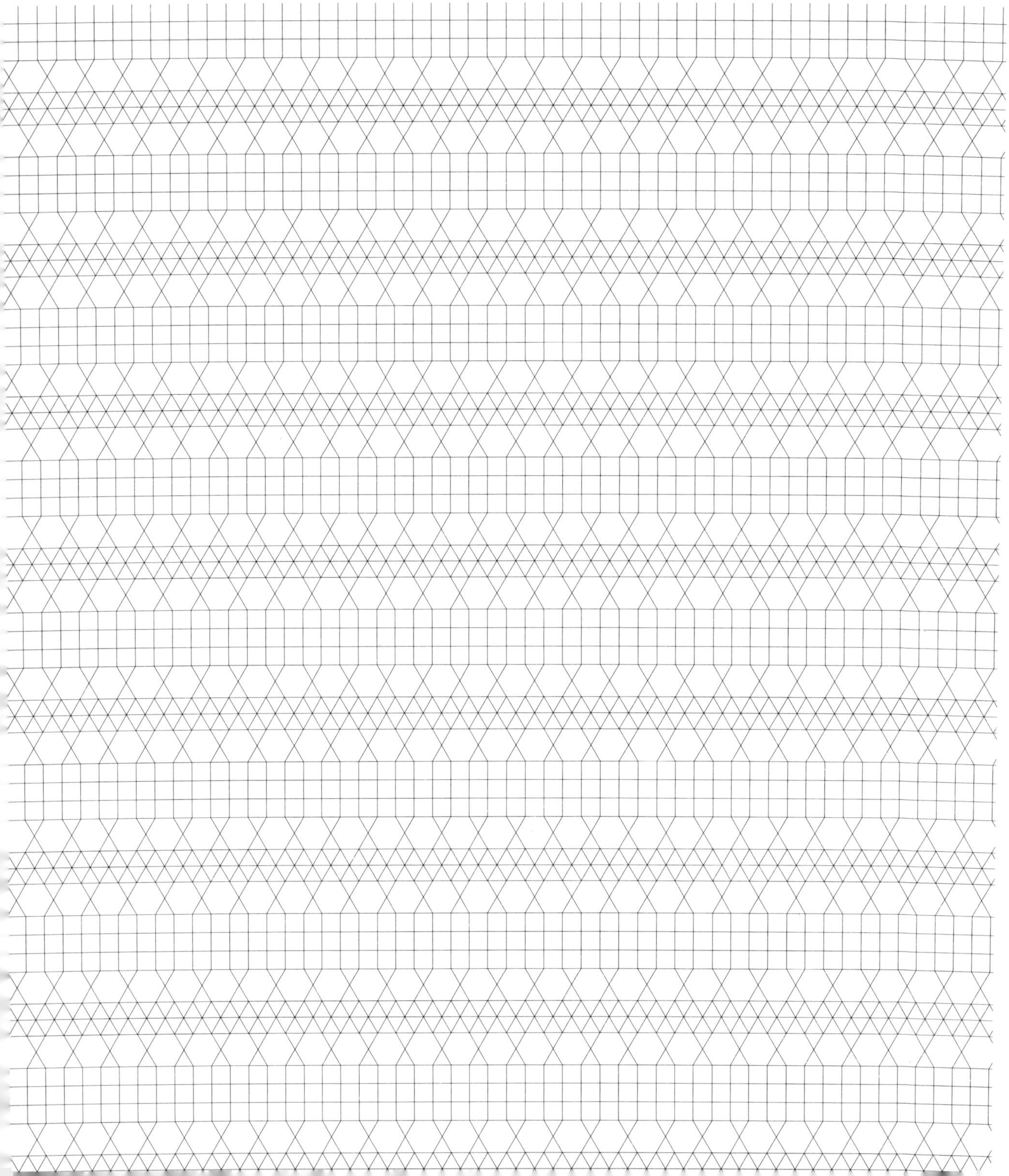

k=5 uniform tiling $\{3^6; 4^4; 3{\cdot}6{\cdot}3{\cdot}6; 3^4{\cdot}6; 3{\cdot}4^2{\cdot}6\}_3$

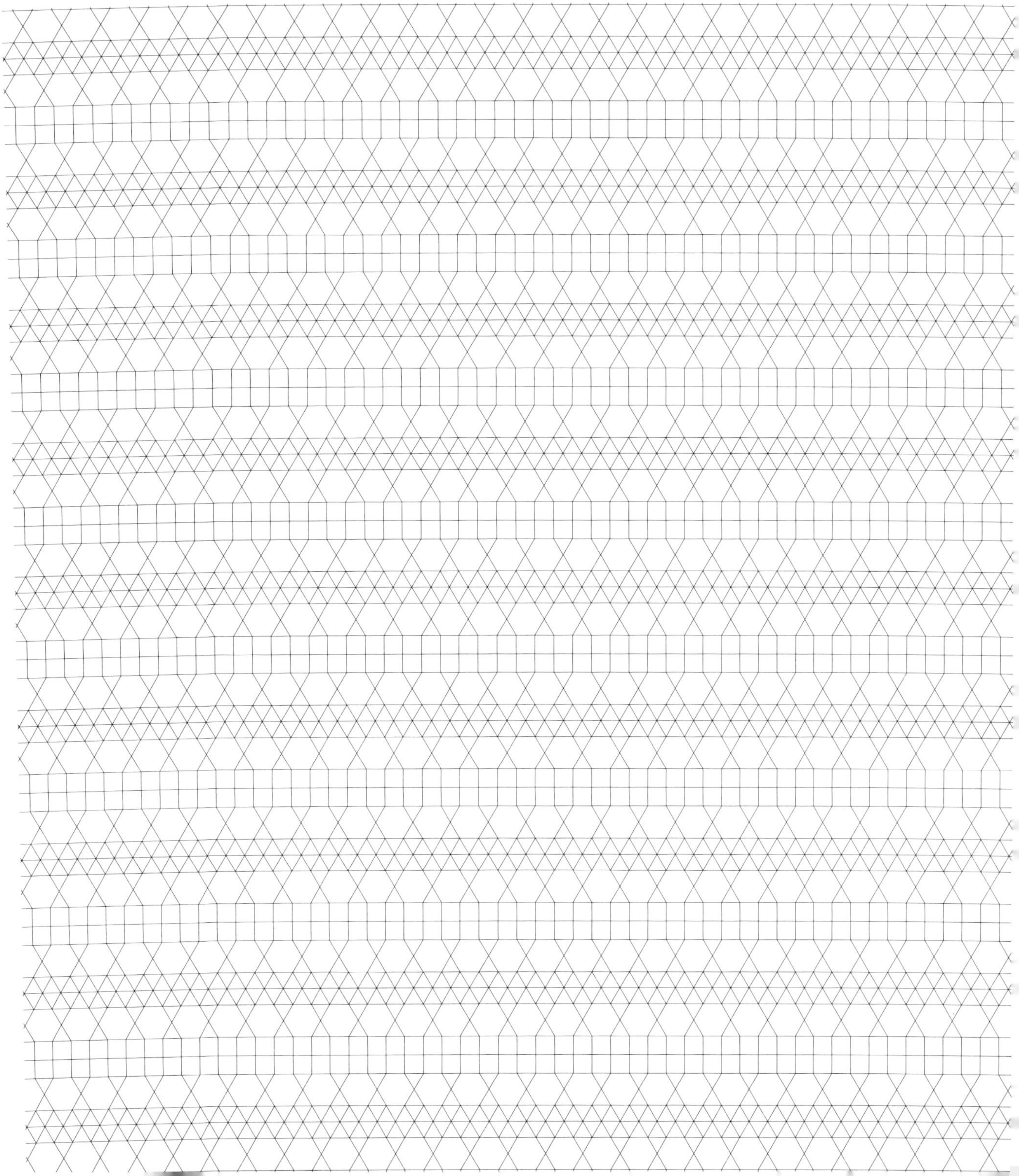

k=5 uniform tiling $\{3^6; 4^4; 3{\cdot}6{\cdot}3{\cdot}6; 3^4{\cdot}6; 3{\cdot}4^2{\cdot}6\}_4$

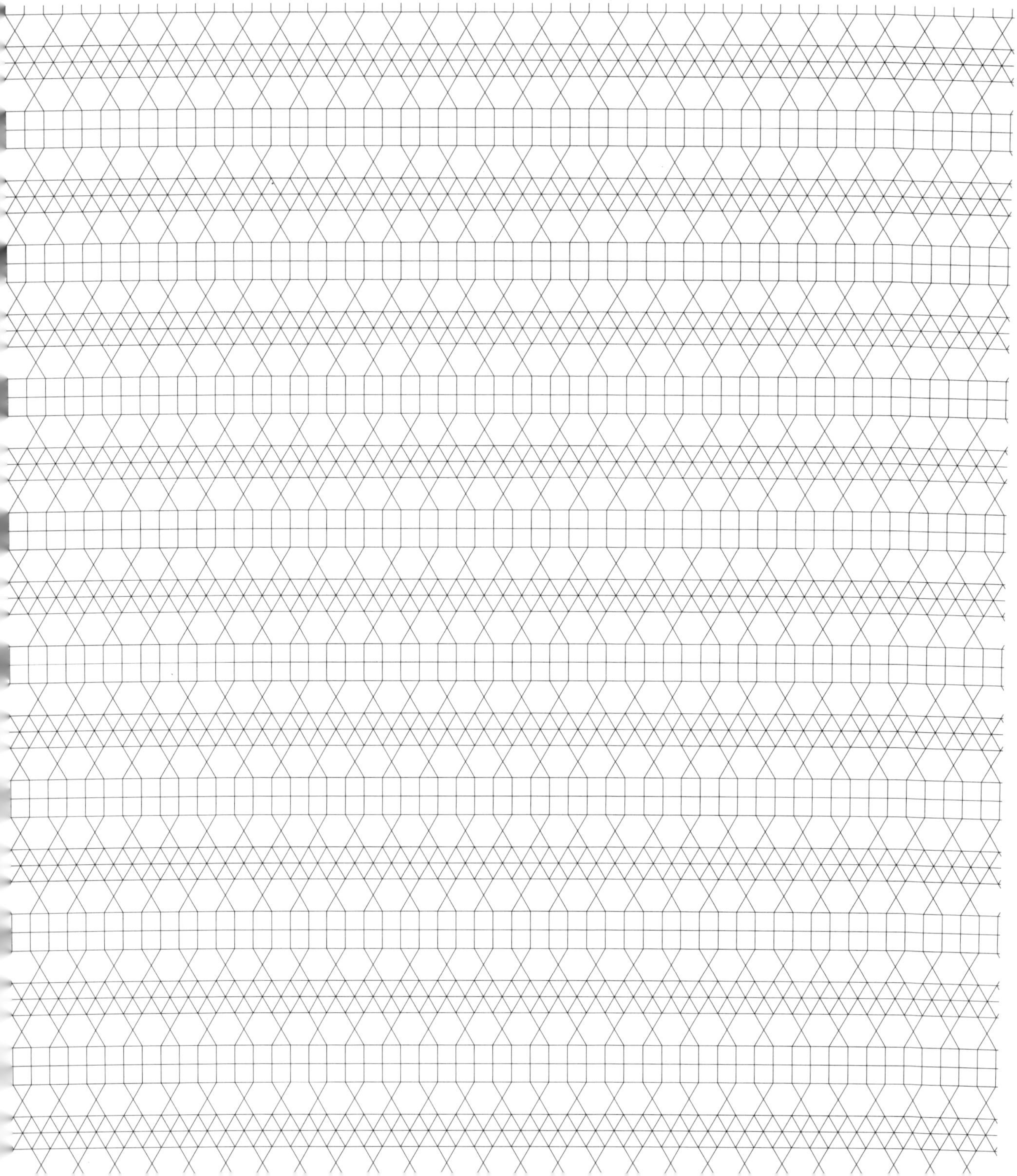

k=5 uniform tiling $\{3^6; 4^4; 3{\cdot}6{\cdot}3{\cdot}6; 3^4{\cdot}6; 3{\cdot}4^2{\cdot}6\}_5$

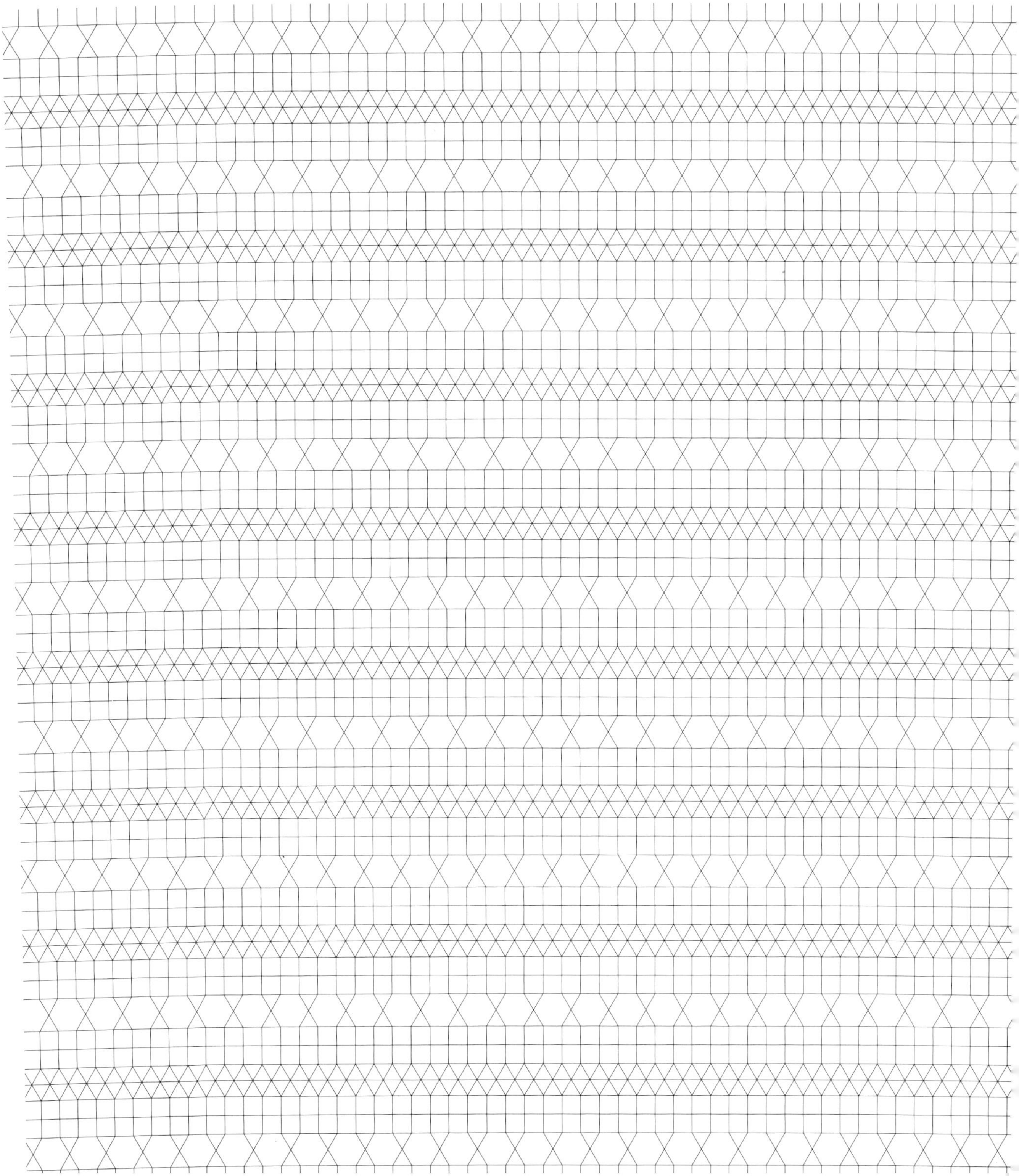

k=5 uniform tiling $\{3^6; 4^4; 3^3{\cdot}4^2; 3^4{\cdot}6; 3{\cdot}4^2{\cdot}6\}$

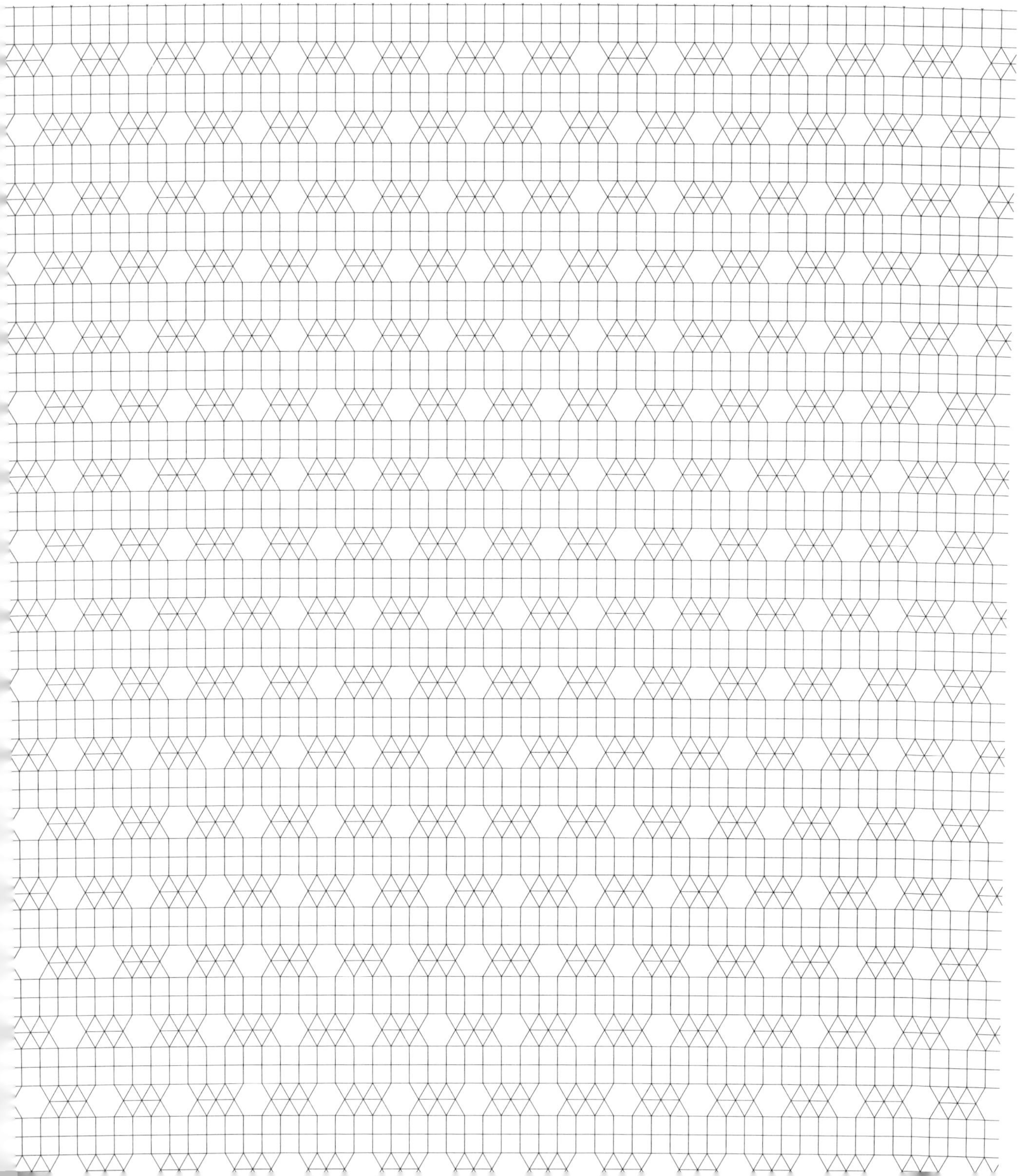

k=5 uniform tiling {3^6; $3^3{\cdot}4^2$; $3^2{\cdot}6^2$; $3{\cdot}6{\cdot}3{\cdot}6$; $3{\cdot}4^2{\cdot}6$}

k=5 uniform tiling $\{3^6; 3^3 \cdot 4^2; 3^2 \cdot 6^2; 3^4 \cdot 6; 3 \cdot 4^2 \cdot 6\}$

k=5 uniform tiling $\{4^4; 3^3{\cdot}4^2; 3^2{\cdot}4{\cdot}3{\cdot}4; 3^4{\cdot}6; 3{\cdot}4^2{\cdot}6\}$

 k=5 uniform tiling $\{3^6; 3^2{\cdot}4{\cdot}3{\cdot}4; 3{\cdot}6{\cdot}3{\cdot}6; 3{\cdot}4^2{\cdot}6; 3{\cdot}4{\cdot}6{\cdot}4\}$

k=5 uniform tiling {3^6; 3·6·3·6; 3^4·6; 3·4^2·6; 3·4·6·4}

k=5 uniform tiling {6^3; $3^2{\cdot}4{\cdot}3{\cdot}4$; $3^2{\cdot}6^2$; $3{\cdot}4^2{\cdot}6$; $3{\cdot}4{\cdot}6{\cdot}4$}

k=5 uniform tiling $\{3^6; 3^3{\cdot}4^2; 3^4{\cdot}6; 3{\cdot}4^2{\cdot}6; 4{\cdot}6{\cdot}12\}$

k=5 uniform tiling

$\{3^2{\cdot}4{\cdot}3{\cdot}4;\ 3{\cdot}12^2;\ 3^2{\cdot}4{\cdot}12;\ 3{\cdot}4{\cdot}6{\cdot}4;\ 4{\cdot}6{\cdot}12\}$

k=6 uniform tiling $\{3^6; 4^4; 3^3{\cdot}4^2; 3^2{\cdot}6^2; 3{\cdot}6{\cdot}3{\cdot}6; 3{\cdot}4^2{\cdot}6\}$

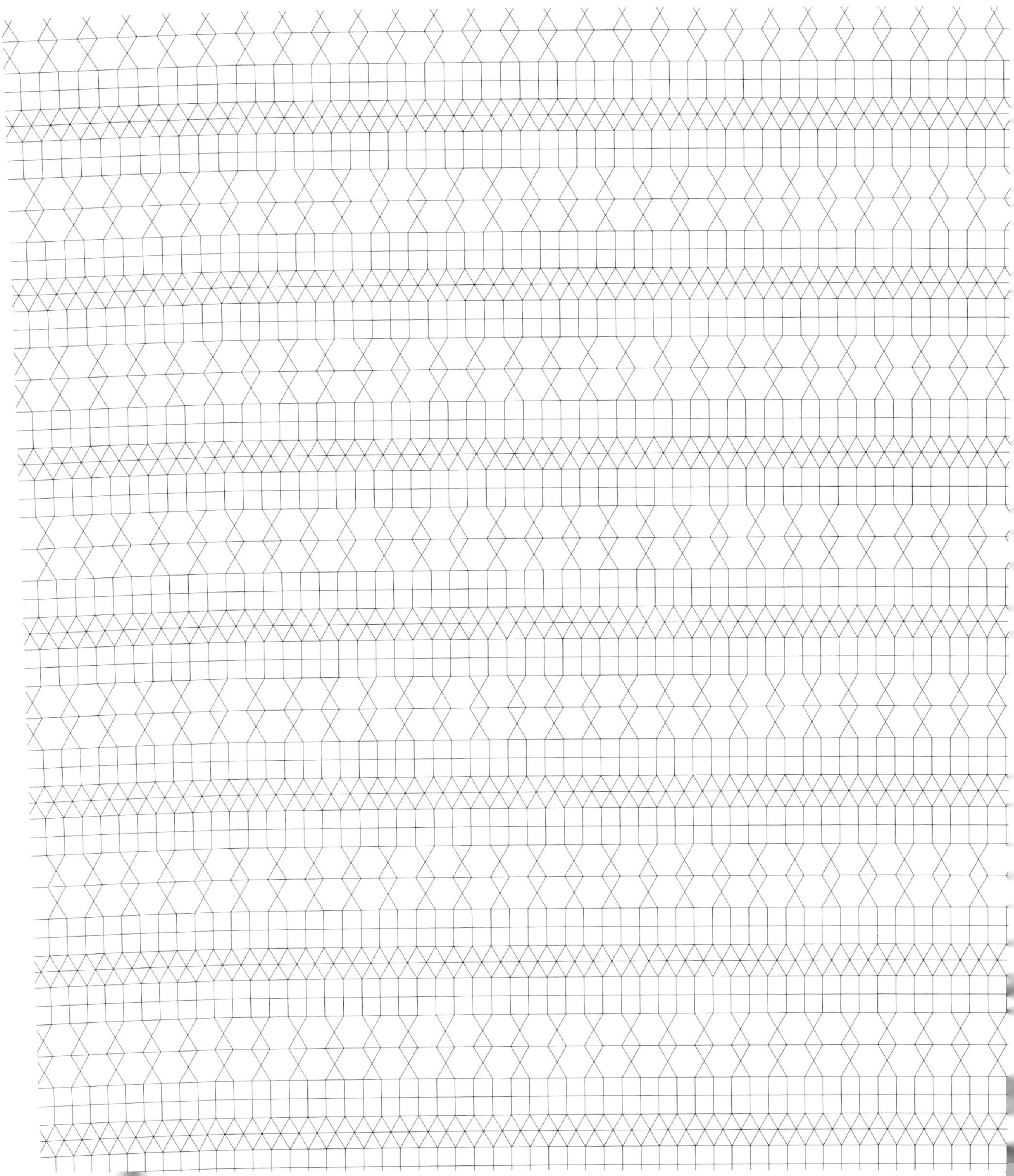

k=6 uniform tiling

$\{3^6; 3^3{\cdot}4^2; 3^2{\cdot}4{\cdot}3{\cdot}4; 3^2{\cdot}6^2; 3{\cdot}4^2{\cdot}6; 3{\cdot}4{\cdot}6{\cdot}4\}$

k=6 uniform tiling {$3^3{\cdot}4^2$; $3^2{\cdot}4{\cdot}3{\cdot}4$; $3^2{\cdot}6^2$; $3^4{\cdot}6$; $3{\cdot}4^2{\cdot}6$; $3{\cdot}4{\cdot}6{\cdot}4$}

k=6 uniform tiling $\{3^6; 3^3{\cdot}4^2; 3^2{\cdot}4{\cdot}3{\cdot}4; 3^2{\cdot}6^2; 3^4{\cdot}6; 3{\cdot}4^2{\cdot}6\}$

k=6 uniform tiling

$\{3^3{\cdot}4^2;\ 3^2{\cdot}4{\cdot}3{\cdot}4;\ 3{\cdot}6{\cdot}3{\cdot}6;\ 3^4{\cdot}6;\ 3{\cdot}4^2{\cdot}6;\ 3{\cdot}4{\cdot}6{\cdot}4\}$

k=6 uniform tiling {4^4; 6^3; 3^3·4^2; 3^2·4·3·4; 3^2·6^2; 3·4^2·6}

k=6 uniform tiling $\{3^3{\cdot}4^2; 3^2{\cdot}4{\cdot}3{\cdot}4; 3^2{\cdot}6^2; 3^4{\cdot}6; 3{\cdot}4^2{\cdot}6; 4{\cdot}6{\cdot}12\}$

k=6 uniform tiling {4^4; $3^3{\cdot}4^2$; $3{\cdot}6{\cdot}3{\cdot}6$; $3{\cdot}4^2{\cdot}6$; $3{\cdot}4{\cdot}6{\cdot}4$; $3^2{\cdot}4{\cdot}12$}

gi-dt-117 k=6 uniform tiling $\{3^6; 3^3{\cdot}4^2; 3{\cdot}6{\cdot}3{\cdot}6; 3^4{\cdot}6; 3{\cdot}4^2{\cdot}6; 3^2{\cdot}4{\cdot}12\}$

k=7 uniform tiling $\{3^6; 4^4; 3^3{\cdot}4^2; 3{\cdot}6{\cdot}3{\cdot}6; 3^4{\cdot}6; 3{\cdot}4^2{\cdot}6; 3{\cdot}4{\cdot}6{\cdot}4\}$

k=7 uniform tiling $\{3^6; 3^3{\cdot}4^2; 3^2{\cdot}4{\cdot}3{\cdot}4; 3{\cdot}6{\cdot}3{\cdot}6; 3^4{\cdot}6; 3{\cdot}4^2{\cdot}6; 3{\cdot}4{\cdot}6{\cdot}4\}_1$

k=7 uniform tiling $\{3^6; 3^3{\cdot}4^2; 3^2{\cdot}4{\cdot}3{\cdot}4; 3{\cdot}6{\cdot}3{\cdot}6; 3^4{\cdot}6; 3{\cdot}4^2{\cdot}6; 3{\cdot}4{\cdot}6{\cdot}4\}_2$

k=7 uniform tiling {3^6; 4^4; 6^3; $3^3·4^2$; $3^2·4·3·4$; $3^2·6^2$; $3·4^2·6$}

k=7 uniform tiling

$\{3^6; 3^3{\cdot}4^2; 3^2{\cdot}4{\cdot}3{\cdot}4; 3^4{\cdot}6; 3{\cdot}4{\cdot}6{\cdot}4; 3^2{\cdot}4{\cdot}12; 4{\cdot}6{\cdot}12\}$

k=7 uniform tiling

$\{3^6; 3^3{\cdot}4^2; 3{\cdot}6{\cdot}3{\cdot}6; 3^4{\cdot}6; 3{\cdot}4^2{\cdot}6; 3{\cdot}4{\cdot}6{\cdot}4; 3^2{\cdot}4{\cdot}12\}$

k=7 uniform tiling {$3^3{\cdot}4^2$; $3^2{\cdot}4{\cdot}3{\cdot}4$; $3^2{\cdot}6^2$; $3{\cdot}4^2{\cdot}6$; $3^2{\cdot}4{\cdot}12$; $3{\cdot}4{\cdot}3{\cdot}12$; $4{\cdot}6{\cdot}12$}

quasiperiodic tiling

3-fold symmetry
6-fold symmetry

quasiregular rhombic tiling

quasiperiodic tiling 5-fold symmetry $\Gamma = 1/6 \cup \Gamma = 1/5$

quasiperiodic tiling 5-fold symmetry $\Gamma = 2/9 \cup \Gamma = 1/4$

 quasiperiodic tiling 5-fold symmetry $\Gamma = 2/7 \cup \Gamma = 1/3$

gi-qt-009 quasiperiodic tiling 5-fold symmetry $\Gamma = 1/3 \cup \Gamma = 3/8$

 quasiperiodic tiling 5-fold symmetry $\Gamma = 3/8 \cup \Gamma = 2/5$

gi-qt-013 quasiperiodic tiling 7-fold symmetry $\Gamma = 3/7 \cup \Gamma = 4/9$

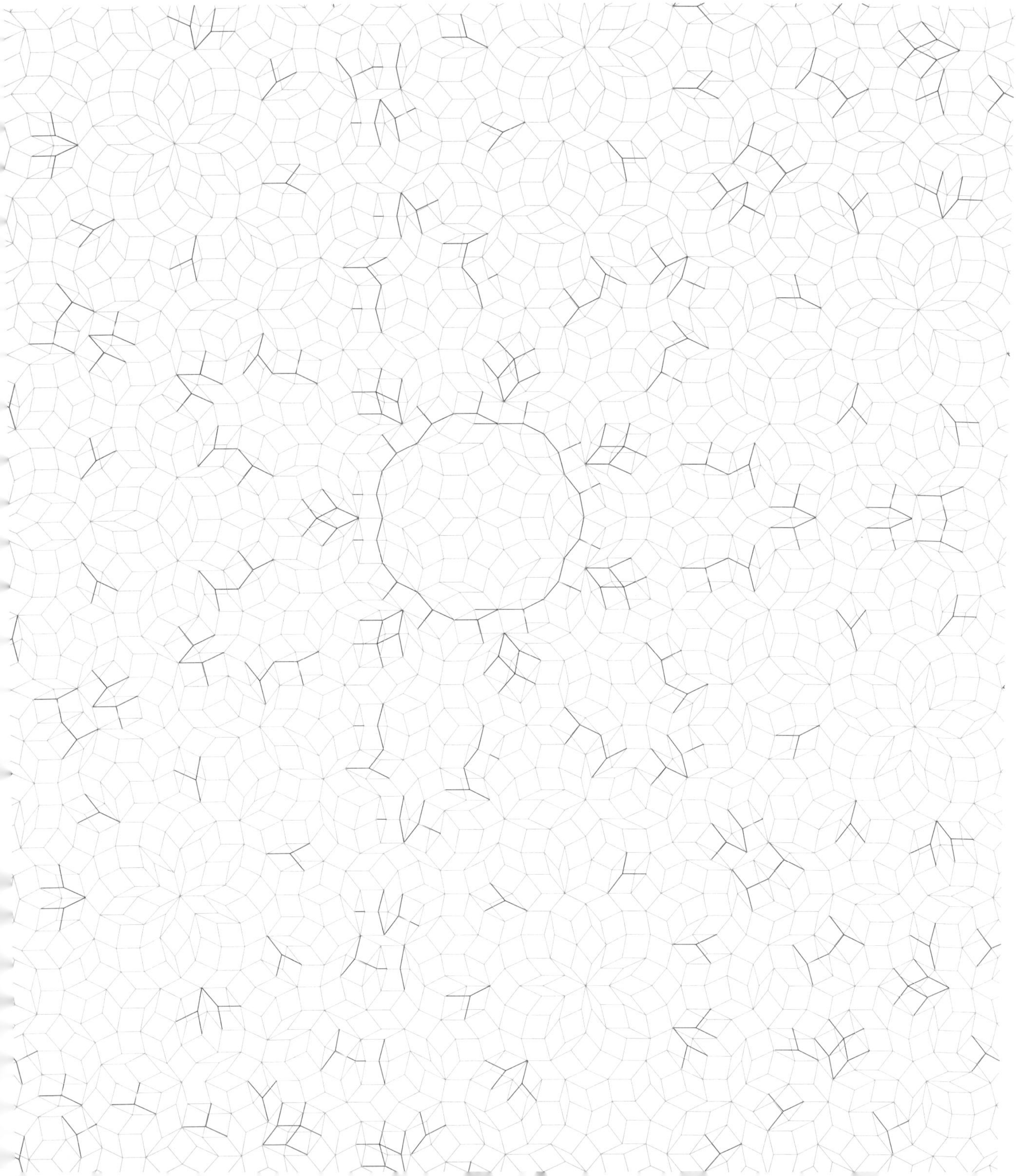

quasiperiodic tiling **7-fold symmetry** $\Gamma = 1/5 \cup \Gamma = 2/9$

quasiperiodic tiling 7-fold symmetry $\Gamma = 1/7 \cup \Gamma = 1/6$

$\Gamma = 1/9 \cup \Gamma = 1/8$

 quasiperiodic tiling 9-fold symmetry $\Gamma = \frac{1}{6} \cup \Gamma = \frac{1}{5}$

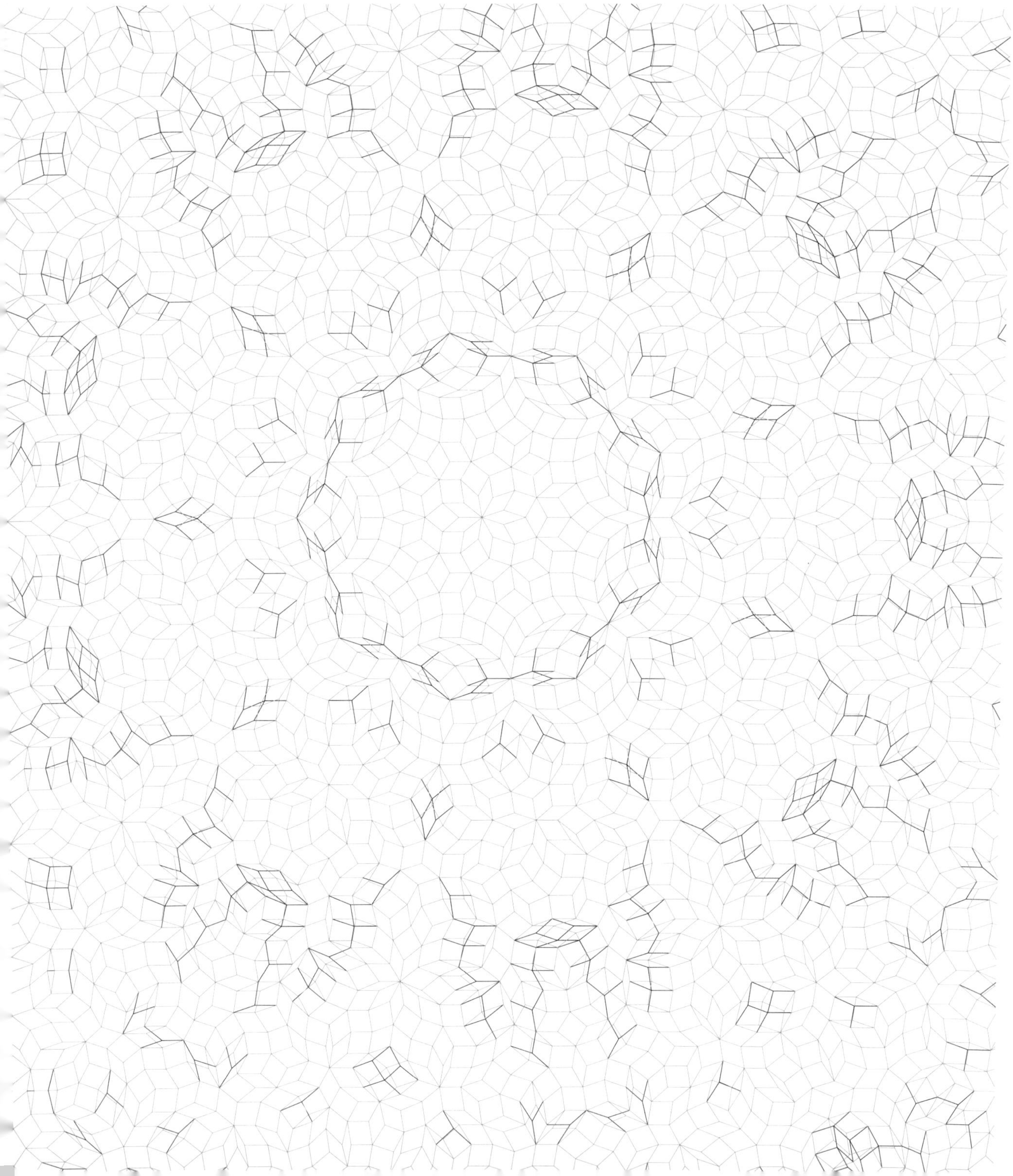

quasiperiodic tiling
union

12-fold symmetry

socolar tiling
socolar tiling inflation

quasiperiodic tiling
union

12-fold symmetry

socolar tiling
shield tiling

quasiperiodic tiling
union

12-fold symmetry

socolar tiling
plate tiling

nonperiodic tiling	penrose kite and dart tiling	0° offset
union	penrose kite and dart tiling	36° offset
	penrose kite and dart tiling	72° offset

meyer construction of a quasicrystal
spiky decagon tiling$_1$

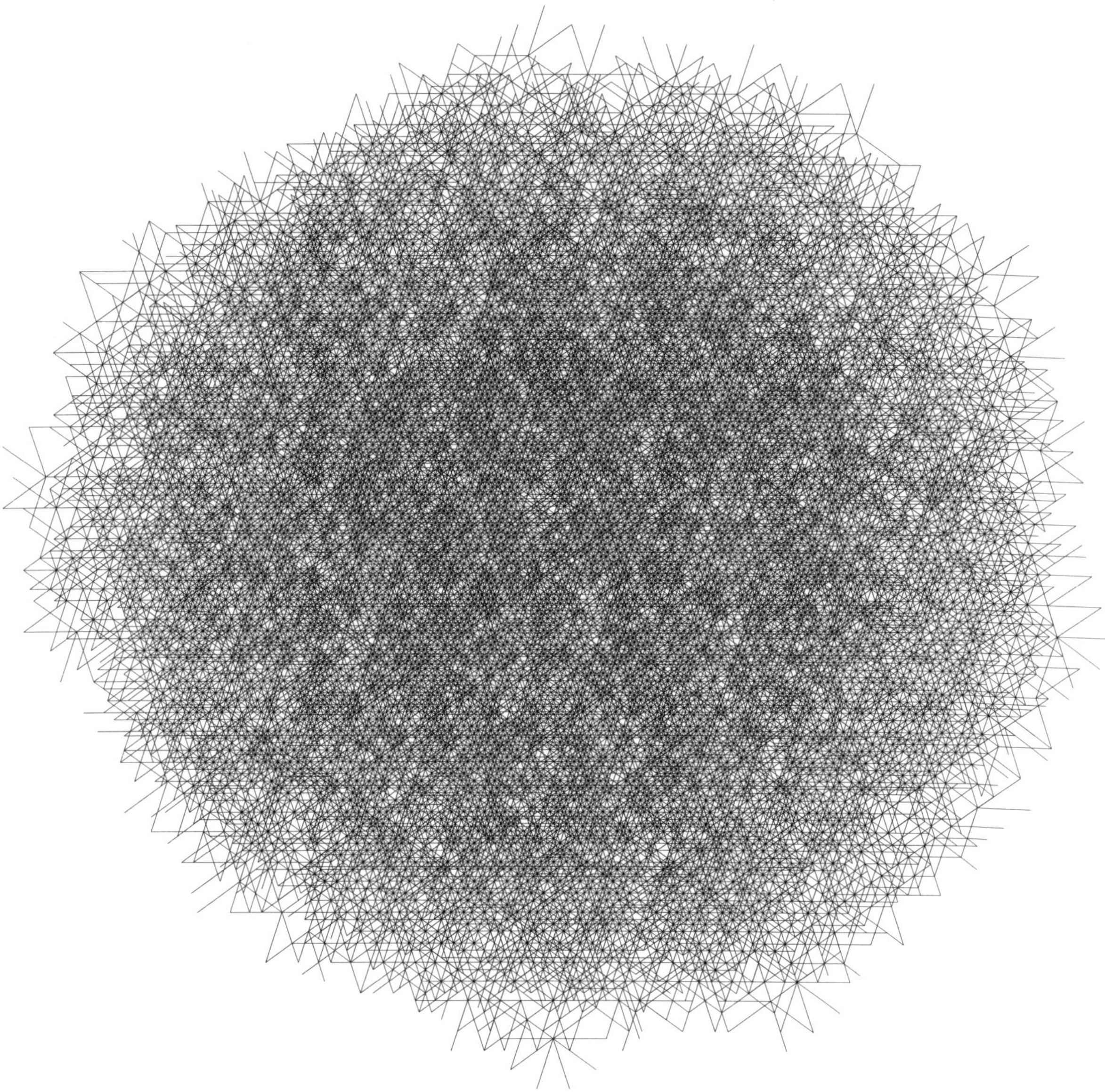

meyer construction of a quasicrystal
spiky decagon tiling$_2$

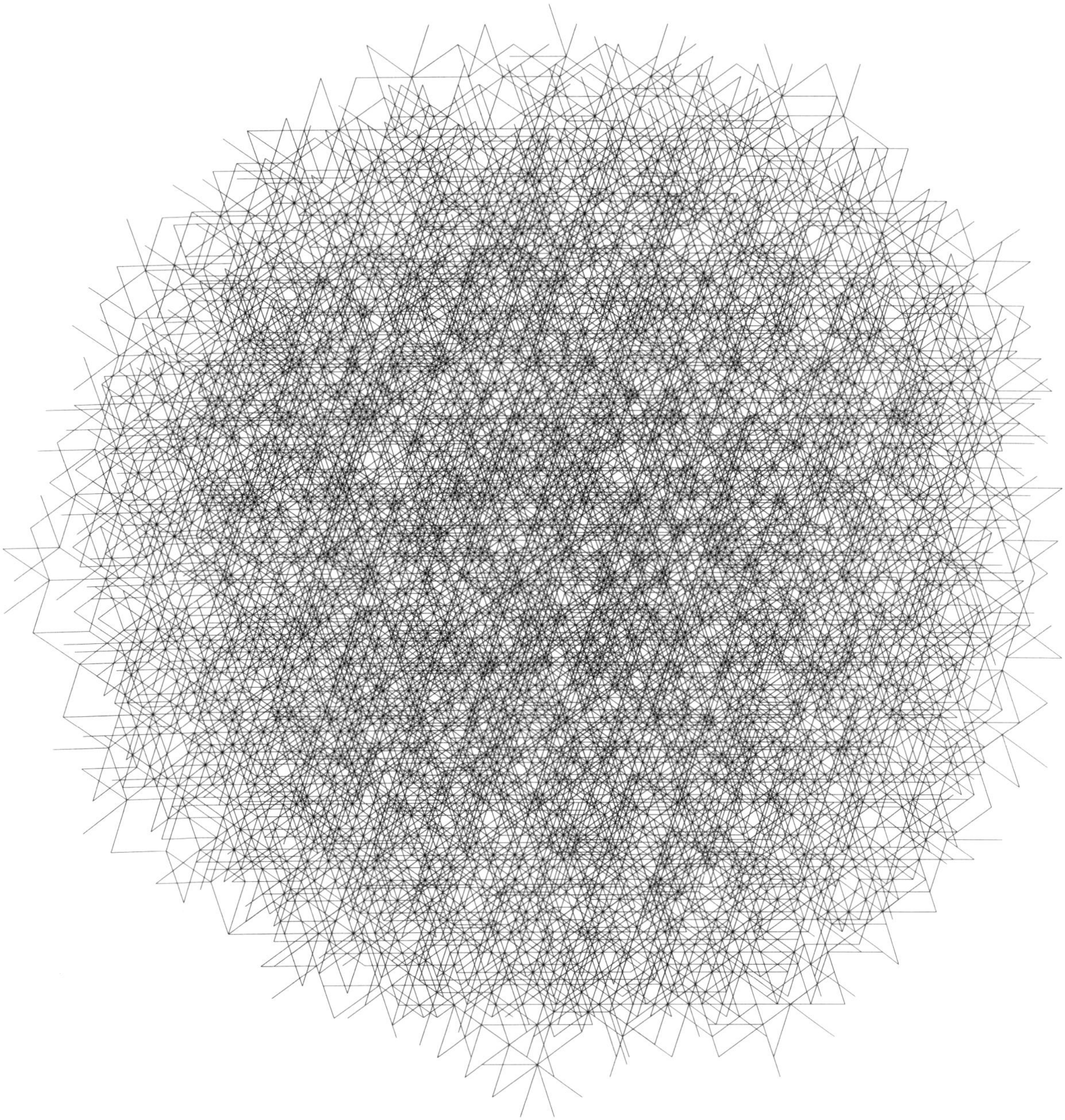

meyer construction of a quasicrystal
spiky decagon tiling$_3$

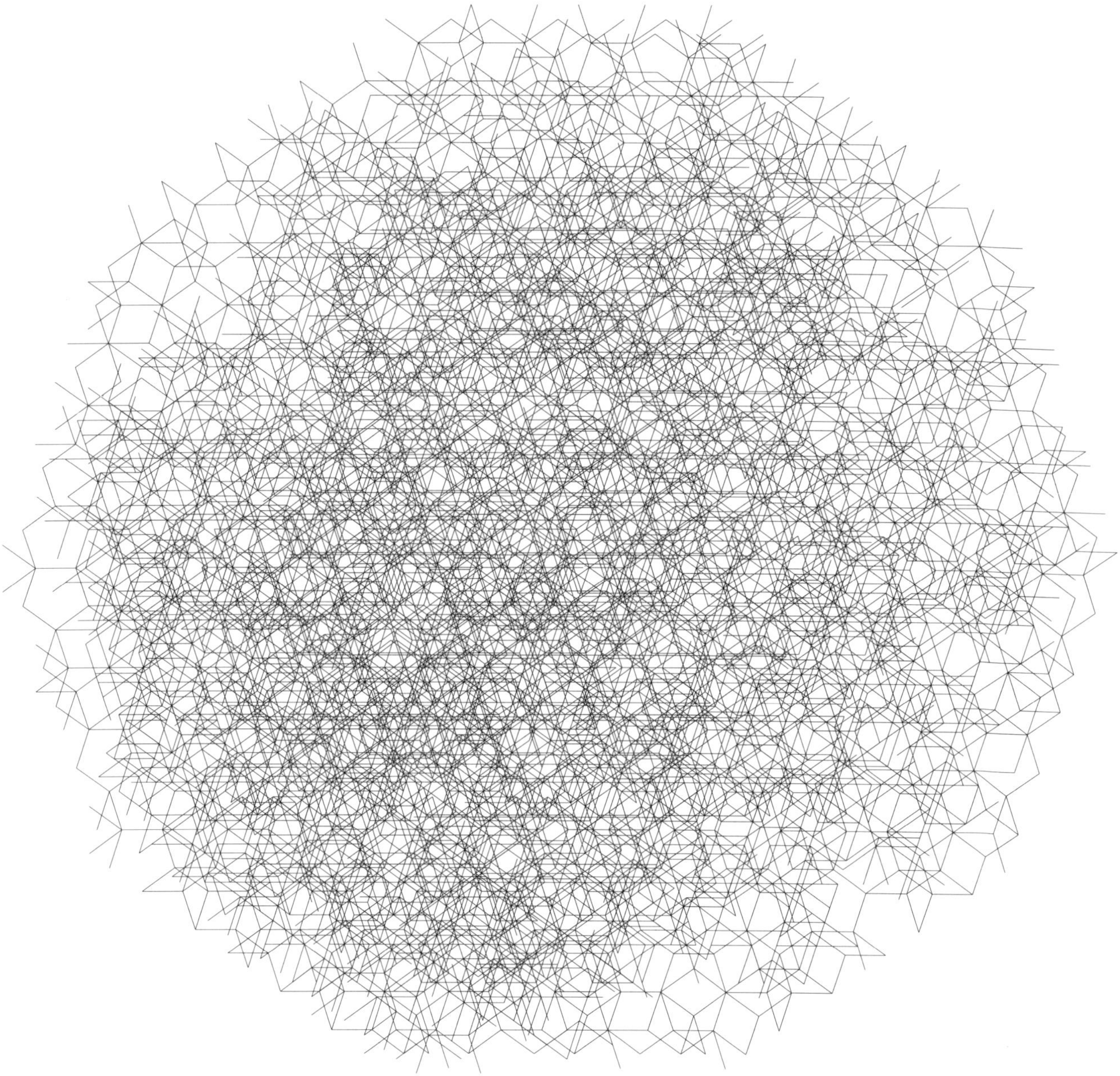

meyer construction of a quasicrystal
spiky decagon tiling$_4$

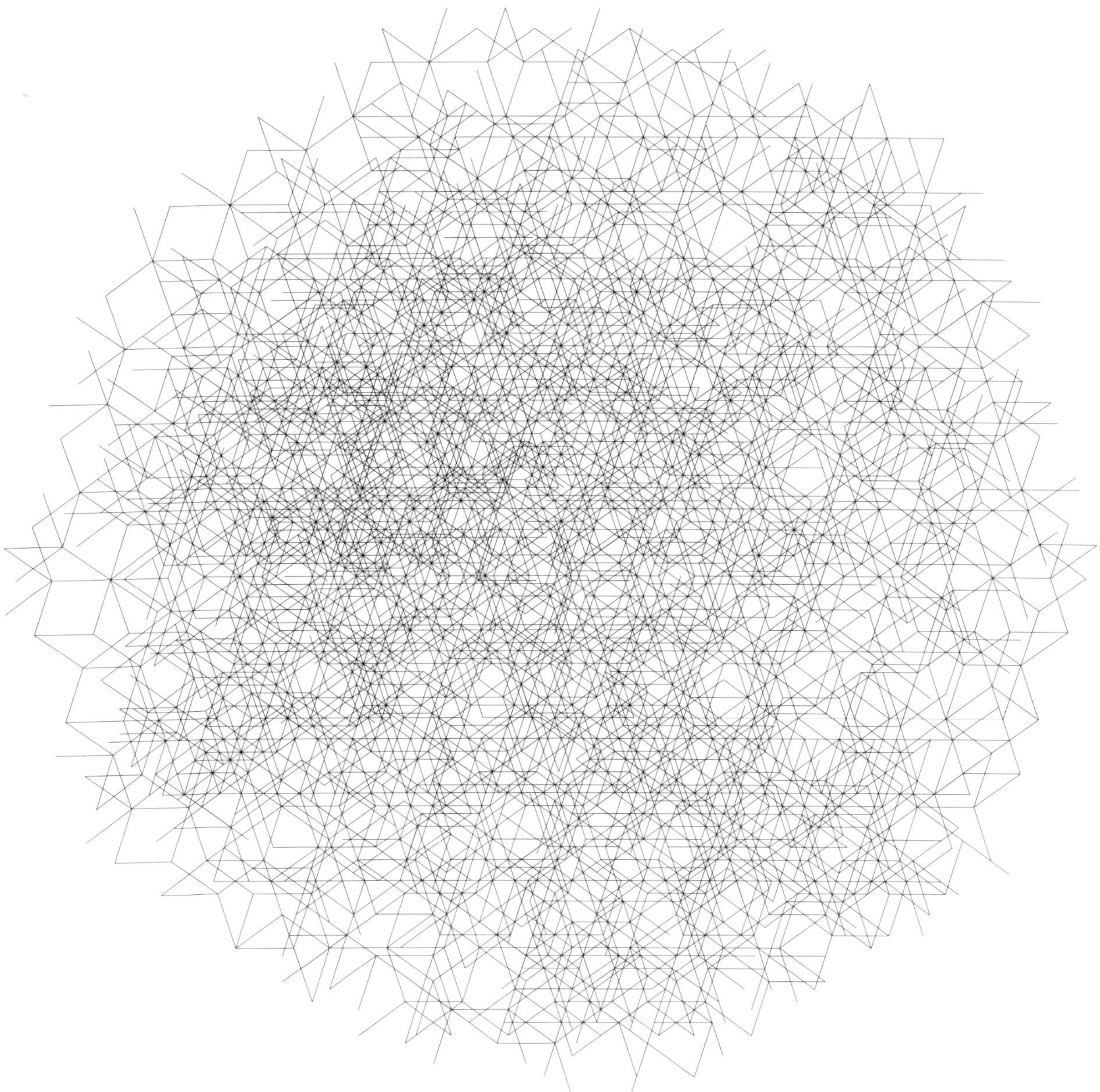

meyer construction of a quasicrystal
spiky decagon tiling$_5$

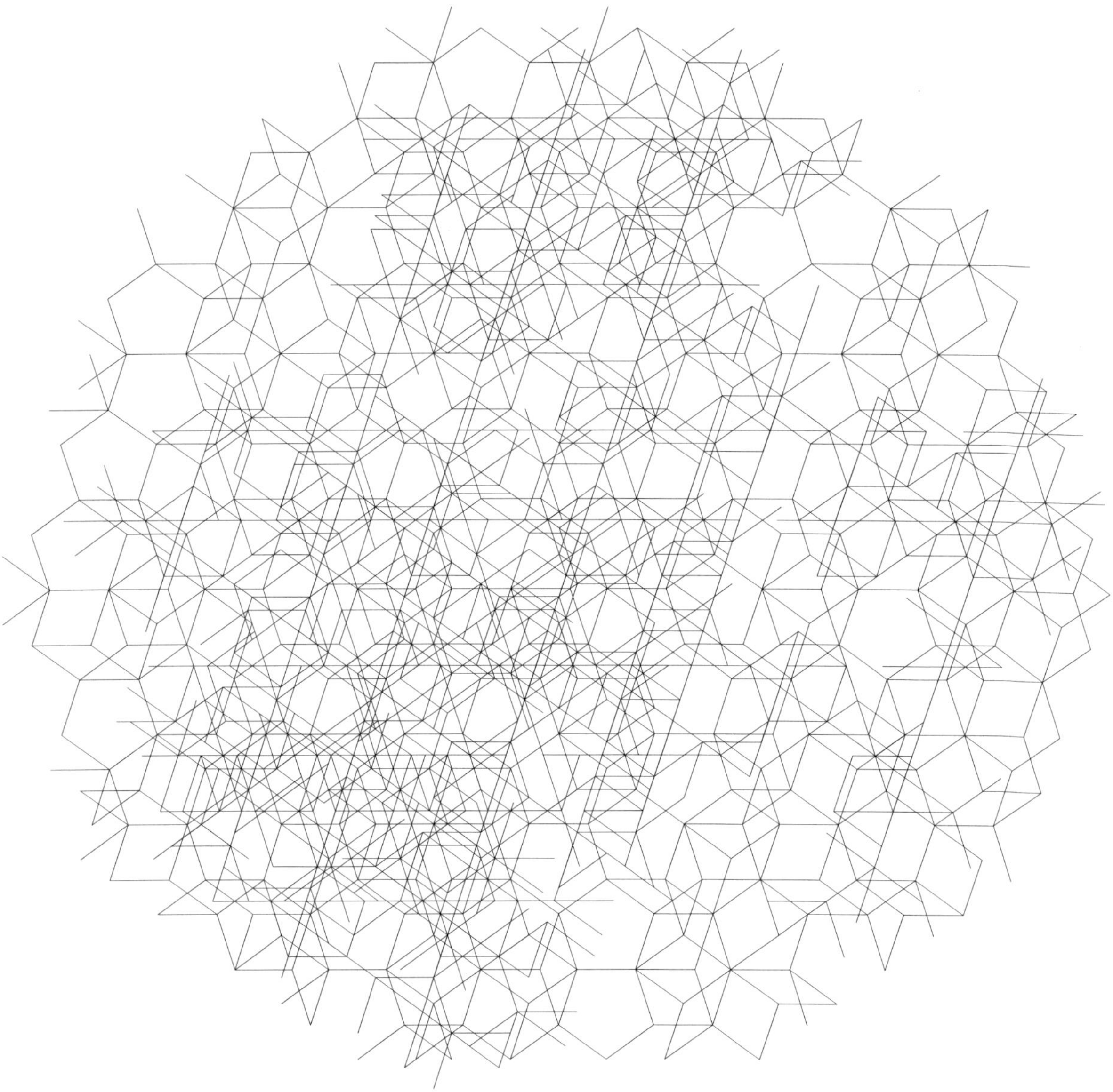

meyer construction of a quasicrystal
spiky decagon tiling$_6$

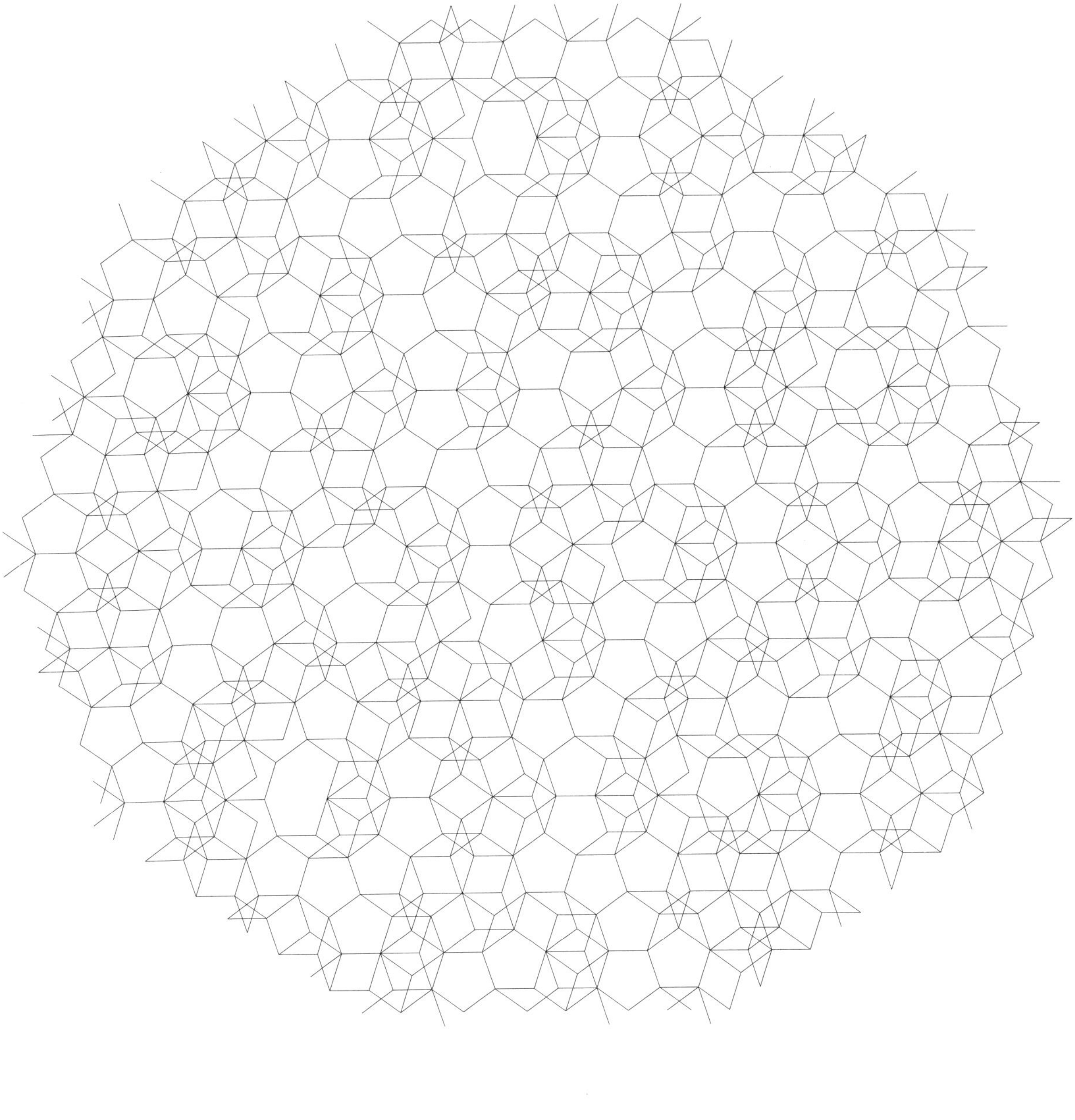

meyer construction of a quasicrystal
spiky decagon tiling$_7$

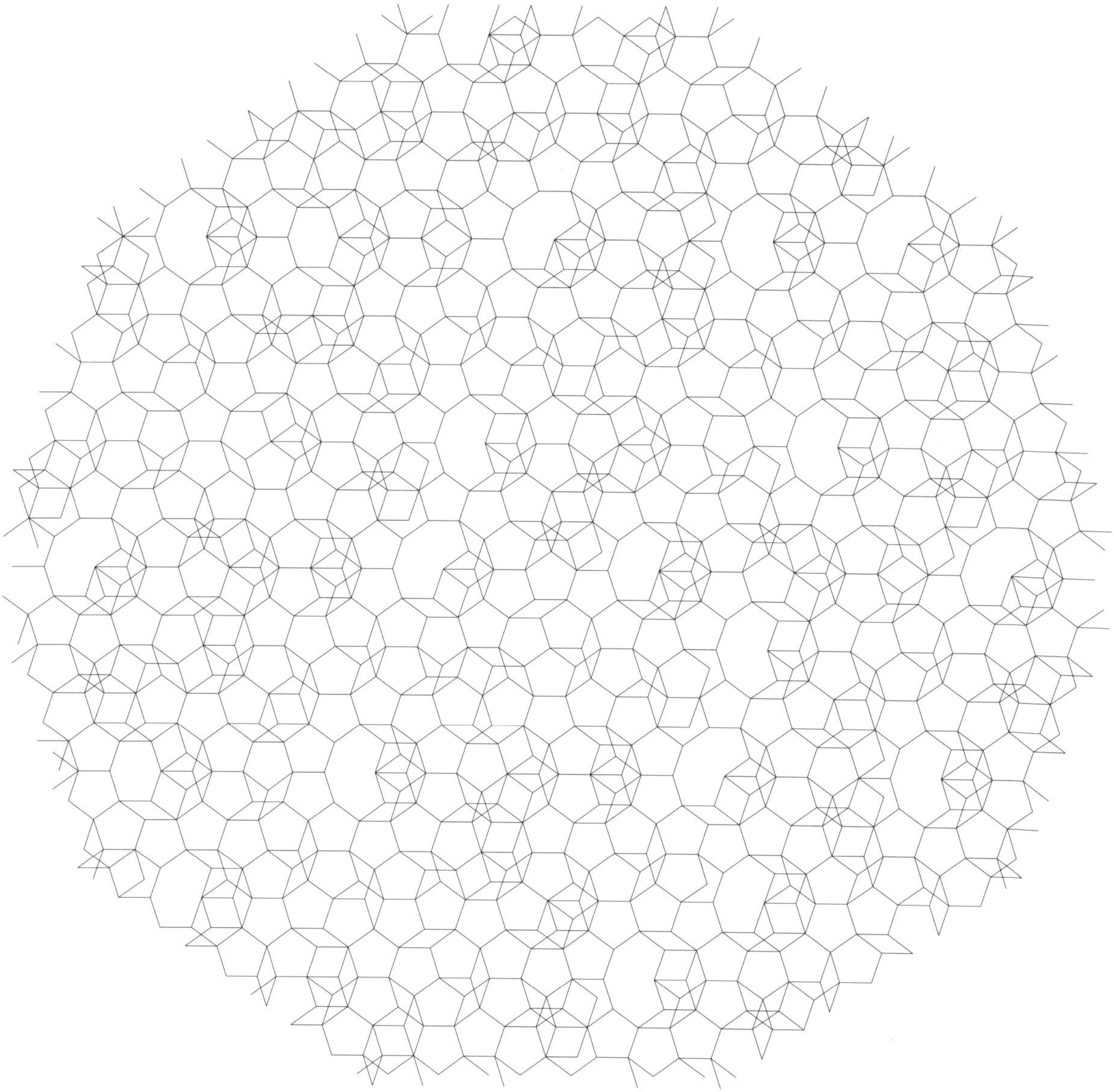

meyer construction of a quasicrystal
spiky decagon tiling$_8$

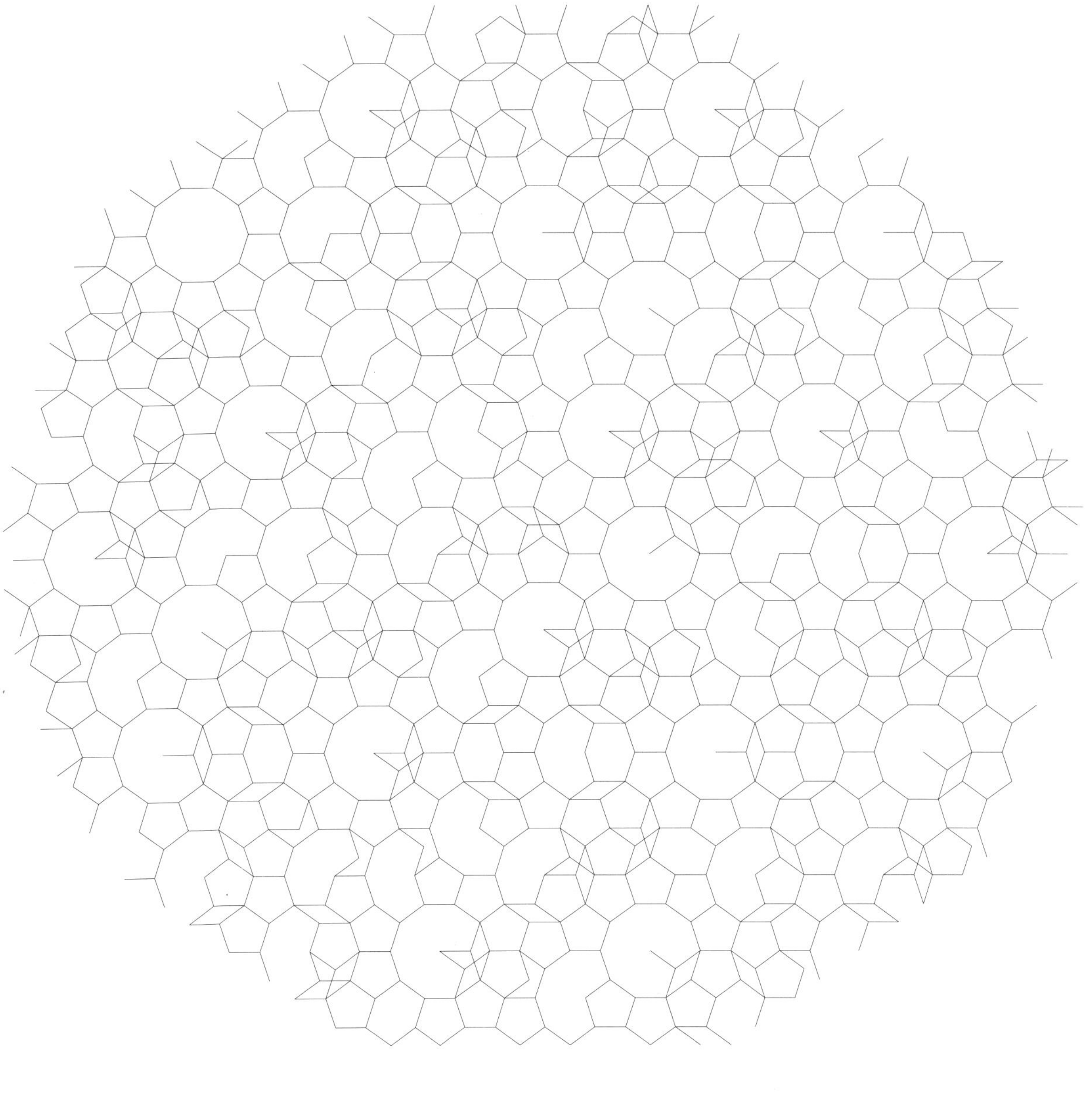

meyer construction of a quasicrystal
spiky decagon tiling$_{10}$

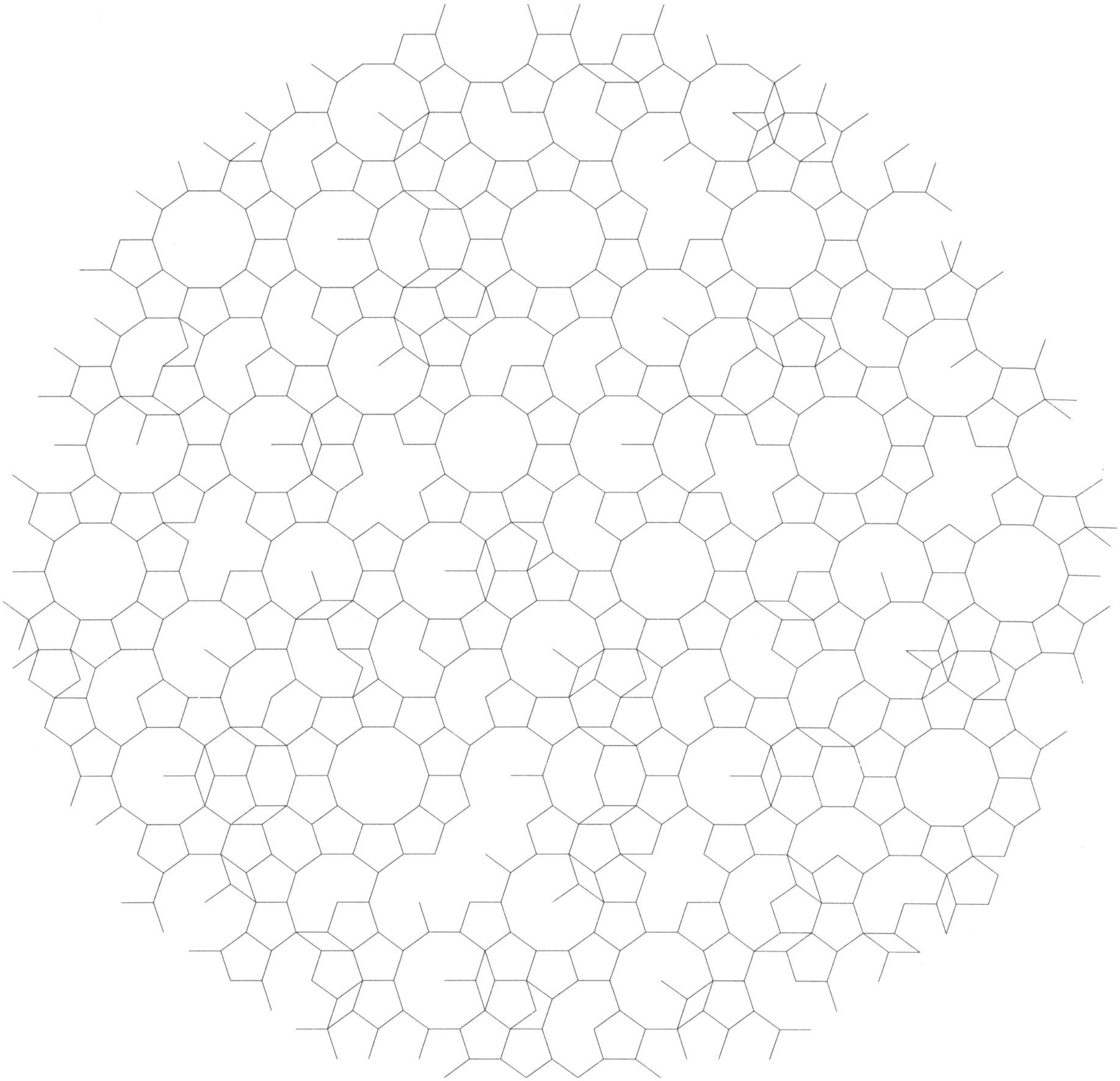

meyer construction of a quasicrystal
spiky decagon tiling$_{11}$

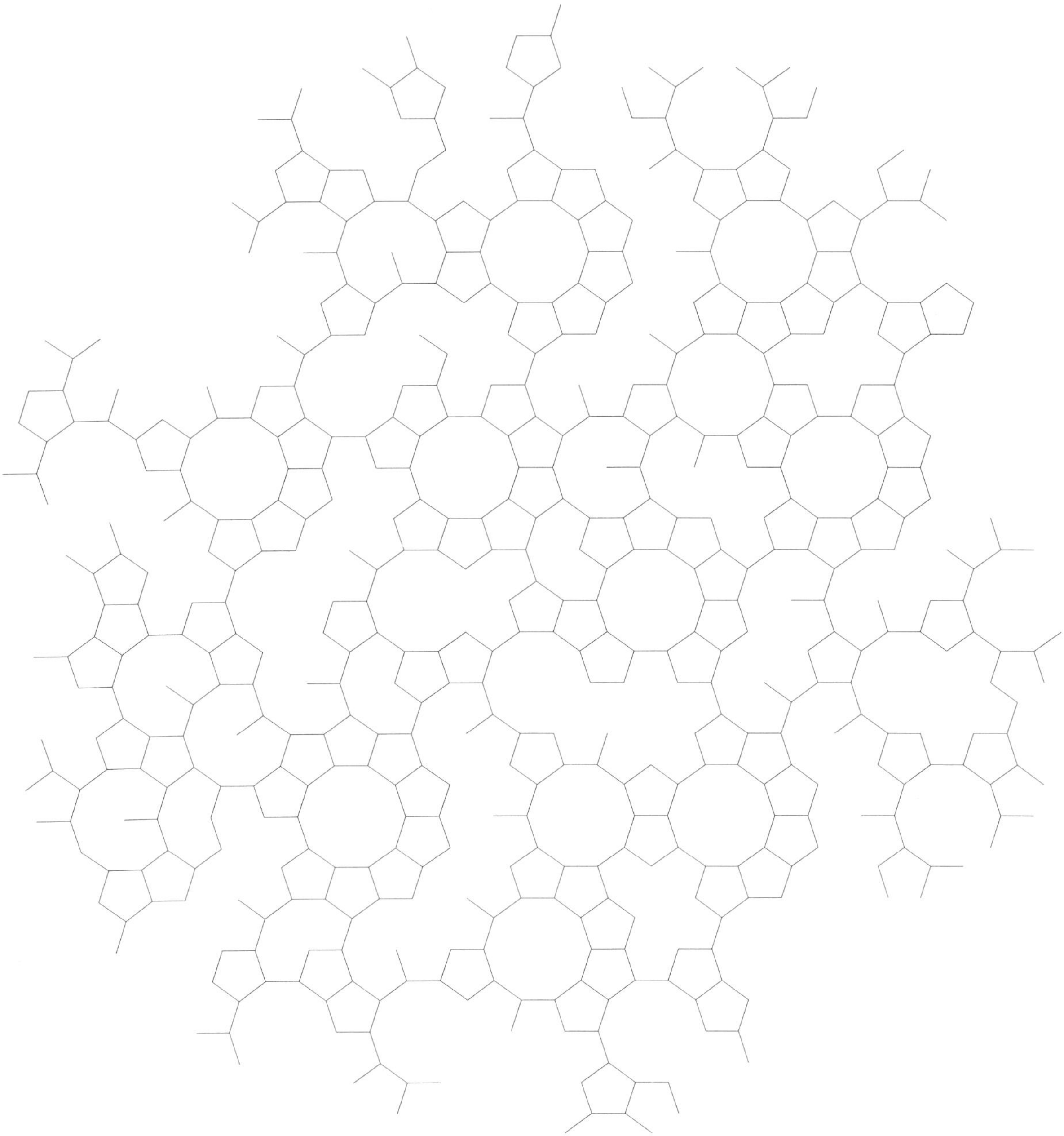

we would like to thank every researcher that made this book possible.
their research and discussion were fundamental in paving the way for the development of our own path; listed here alphabetically are the names of every author whose articles were used throughout the construction of grid index:

robert ammann · helmer aslaksen · michael baake · f. beenker · shelomo i. ben-abraham · donald l. d. caspar · nicolaas de bruijn · albrecht dürer · eric fontano · dirk frettlöh · franz gähler · brian galebach · uwe grimm · branko grünbaum · petra gummelt · istván hargittai · edmund harriss · christian janot · jay kappraff · johannes kepler · vladimir a. koptsik · otto krotenheerdt · alexandre losev · reinhard lück · roger penrose · john savard · radmila sazdanovic · michael schreiber · marjorie senechal · george c. shepherd · aleksei v. shubnikov · miodrag sremcevic · srdjan vukmirovic _publications

GRÜNBAUM, Branko, SHEPHERD, George C.
Tilings and Patterns: An Introduction, New York: W. H. Freeman and Co., 1989

HARGITTAI, István (ed.)
Fivefold Symmetry, Singapore: World Scientific, 1992

JANOT, Christian
Quasicrystals: A Primer, Oxford: Oxford Science Publications, 1995

KAPPRAFF, Jay (ed.)
Connections: The Geometric Bridge Between Art and Science, Singapore: World Scientific, 2001

SHUBNIKOV, Aleksei V., KOPTSIK, Vladimir A.
Symmetry in Science and Art, New York/ London: Plenum Press, 1974

_articles

ASLAKSEN, Helmer, "In Search of Demiregular Tilings"
helmer aslaksen: list of publications
http://www.math.nus.edu.sg/aslaksen/helmerpub.html

BAAKE, Michael, "A Guide to Mathematical Quasicrystals"
research group michael baake: preprints & recent articles
http://www.math.uni-bielefeld.de/baake/preprints.html#

GÄHLER, Franz et al., "Dodecagonal Tilings as Maximal Cluster Coverings"
dr. franz gähler: list of publications
http://www.itap.physik.uni-stuttgart.de/~gaehler/publications.html

GÄHLER, Franz et al., "Generation of Quasiperiodic Order by Maximal Cluster Covering"
http://www.itap.physik.uni-stuttgart.de/~gaehler/publications.html

GÄHLER, Franz, "Cluster Coverings: A Powerful Ordering Principle for Quasicrystals"
http://www.itap.physik.uni-stuttgart.de/~gaehler/publications.html

GUMMELT, Petra, "Random Cluster Covering Model"
homepage - petra gummelt: papers: (pdf, ps)
http://www.math-inf.uni-greifswald.de/~gummelt/pubprep.html

LOSEV, Alexandre, "Wave models of non-crystallographic structures"
cornell university library: arXiv.org
http://arxiv.org/abs/0706.0187v2

SENECHAL, Marjorie, "Tilings, Diffraction, and Quasicrystals"
The Mathematica Journal, vol. 4, issue 2, spring 1994, pp. 10-15

_packages

GRIMM, Uwe, SCHREIBER, Michael, "Aperiodic Tilings on the Computer"
uwe grimm's home page: mathematica programmes on quasiperiodic tilings
http://mcs.open.ac.uk/ugg2/AperiodicTilings

SREMCEVIC, Miodrag et al., "Tessellations of the Euclidean, Elliptic and Hyperbolic Plane"
wolfram library archive: mathsource
http://library.wolfram.com/infocenter/MathSource/4540/

carsten nicolai . *grid index*
edited and preface by carsten nicolai

design: nuno da luz, carsten nicolai
typeface: akzidenz-grotesk pro by berthold typefoundry

project management: elisabeth honerla for gestalten
production management: janni milstrey for gestalten
printer: c&c offset printing, hong kong
printed in china.

published by gestalten, berlin 2009
ISBN 978-3-89955-241-6

2nd printing, 2009

for further information, please visit
www.gestalten.com
www.carstennicolai.de

carsten nicolai is represented by
galerie eigen + art, leipzig/ berlin
pacewildenstein, new york

bibliographic information published by Deutsche Nationalbibliothek.
Deutsche Nationalbibliothek lists this publication in the Deutsche Nationalbibliografie; detailed bibliographic data is available on the internet at http://dnb.d-nb.de

this book was printed according to the internationally accepted ISO 14001 standards for environmental protection, which specify requirements for an environmental management system.
this book is printed on certified paper that comes from sources practicing responsible forest management.

myclimate
Protect our planet

gestalten is a climate neutral company and so are our products. we collaborate with the non-profit carbon offset provider myclimate (www.myclimate.org) to neutralize the company's carbon footprint produced through our worldwide business activities by investing in projects that reduce CO_2 emissions (www.gestalten.com/myclimate).